Benjamin Parke Avery

Californian Pictures in Prose and Verse

Benjamin Parke Avery

Californian Pictures in Prose and Verse

ISBN/EAN: 9783337417697

Printed in Europe, USA, Canada, Australia, Japan

Cover: Foto ©Thomas Meinert / pixelio.de

More available books at **www.hansebooks.com**

CALIFORNIAN PICTURES

IN

PROSE AND VERSE

BY

BENJAMIN PARKE AVERY

NEW YORK
PUBLISHED BY HURD AND HOUGHTON
Cambridge: The Riverside Press
1878

TO THE MEMORY OF

SAMUEL PUTNAM AVERY,

OBIT NEW YORK, 1832.

CONTENTS.

	PAGE
A WORD BEFORE	9
NATURE AND ART	13
A WILD NOSEGAY	18
MOUNTAIN, LAKE, AND VALLEY	19
A MEMORY OF THE SIERRA	63
UP THE WESTERN SLOPE	66
SUNRISE NEAR HENNESS PASS	84
ON THE SUMMIT	87
EL RIO DE LAS PLUMAS	118
HEAD-WATERS OF THE SACRAMENTO	121
THE BIRTH OF BEAUTY	150
ASCENT OF MOUNT SHASTA	152
THE MEADOW LARK	191
THE GEYSERS	193
GOLDEN GATE PARK	236
CITY SCENERY	239

	PAGE
THE FAWN ON CHANGE	261
SANTA CRUZ MOUNTAINS	264
AUTOCHTHONES	280
THE FIRST PEOPLE	282
SONG OF THE VAQUERO	311
THE TRINITY DIAMOND	314
OLD AND NEW	343

ILLUSTRATIONS.

	DRAWN BY	ENGRAVED BY	PAGE
FRONTISPIECE. (From a Sketch by G. A. Frost)	Thomas Moran	W. H. Morse.	
FOOT-HILLS OF THE COAST RANGE. (From a Sketch by W. Keith)	Thomas Moran	J. A. Bogert	24
YOSEMITE FALLS, FROM GLACIER POINT. (From a Photograph)	Alfred Kappes	F. Juengling	44
SECTION OF SNOW-SHED. (From a Photograph)	C. A. Vanderhoof	Robert Varley	91
CROWN OF THE SIERRA. (From a Sketch by W. Keith)	Thomas Moran	E. Bookhout	109
DONNER LAKE. (From a Photograph)	Alfred Kappes	F. Juengling	110
MOUNT SHASTA, FROM CASTLE LAKE. (From a Sketch by H. G. Bloomer)	Thomas Moran	J. S. Harley	122
GOLDEN GATE, FROM CONTRA COSTA HILLS. (From a Sketch by W. Keith)	Thomas Moran	Robert Varley	192
VALLEY OAKS. (From a Photograph)	Alfred Kappes	John P. Davis	198
MOUNT ST. HELENA. (From a Photograph)	Alfred Kappes	Bookhout Bros.	202
MOUNT TAMALPAIS. (From a Sketch by W. Keith)	W. H. Gibson	Horace Baker	246
MOUNT DIABLO. (From a Sketch by W. Keith)	W. H. Gibson	F. S. King	250
LOMA PRIETA. (From a Sketch by W. Keith)	W. H. Gibson	Meeder & Chubb	278

A WORD BEFORE.

The only aim of the following pages is to present a few word-sketches of Californian scenery-studies from nature, true to local color and form, and barely indicating the salient characteristics of plant and animal life and rocky structure. Those who love nature for her own sake, and for her relations to the best art, will sympathize with the motive, whatever may be the imperfection, of these sketches. Some of them originally appeared in the "Overland Monthly," but these have been retouched for this volume. The closing sketches, under the titles of "The First People" and "The Trinity Diamond," are not entirely descriptive of scenery, but introduce the figures of the Indians and the roving miners, who were once very characteristic of Californian landscapes, and still remain a part of them. All of the illustrations are after drawings or photographs from nature, the former made by artists who have found their whole inspiration in California, and who are helping to create there an original school of art. The interspersed verses make no poetical pretensions. They are intended only as pictures in rhyme, and not finished pictures at that.

It should be stated that descriptions of some of the most remarkable scenery in California — such as Yosemite, the Big Tree groves, and those regions of the high sierra lying in the southern part of the State — are purposely withheld, for the reason that they have been already better described by Prof. J. D. Whitney, Clarence King, and John Muir, have been illustrated and written about by scores of artists and authors, and have so become in a measure hackneyed. It was, besides, the wish of the present writer to describe what was most familiar and recent in his own experience. So much admiration has been lavished on two or three grand features of the State, that picturesque details and the claims of less celebrated spots are neglected. Even in this little work much is overlooked that will yet employ profitably abler hands.

CALIFORNIAN PICTURES.

NATURE AND ART.

When Art was young, Pygmalion formed
    A marble maid, divinely fair;
Her beauty all his being warmed,
    And moved him to enraptured prayer:

"Oh, leave her not a senseless stone,
    Almighty Jove, enthroned above!
But give her life to bless my own,
    Endow her with the soul of love!"

Jove heard, and smiled. The marble flushed
    Like snow-peak at the coming sun:
"Pygmalion!" Lo! she spoke and blushed!
    And thus his stainless bride he won.

And ever since the artist-touch
   Has had a quick, Promethean fire,
For all who love their labor much,
   Who nobly struggle and aspire.

To such the miracles recur
   That only genius works at will,
That seem dead images to stir,
   And every source of feeling thrill.

Thus Nature ever, to the heart
   That rightly seeks her, answer gives;
In every master-work of Art
   A portion of her spirit lives.

The templed pile, the marble shape,
   The painted tree, the stream, the sod,
Are only forms her soul to drape —
   For "Nature is the art of God!"

The painter, when he spreads his tints,
   That only mimic what is real,
If Nature guides him, nobly hints
   Her dearest charm, her sweet ideal.

NATURE AND ART.

The rose a richer beauty takes
    From hands that she has deftly taught;
The violet sweeter perfume makes
    When Art has wedded it to thought.

O Goddess! On thy altar tops
    Of awful peaks that touch the blue,
Where every snowy gem that drops
    Unmelted lies in stainless hue,

I gaze upon thy wide domain
    From mountain unto boundless sea,
And listen to the grand refrain
    The pillowed forests sing to thee;

For down below, in circling ranks,
    The pines uplift their branching arms;
And farther, on the river banks,
    The oaks reveal their milder charms.

And as I leave the dizzy height,
    Returning to the valley mead,
Gray rocks with lichens are bedight,
    And flowers up-spring of lowly breed.

The happy creatures of the wild
   Bound from the thicket on my way —
The mother doe, the fawn her child —
   As half in fright and half in play.

By springs where viny tresses cling,
   And tuneful gurgles meet the ear,
The feathered people drink and sing,
   Or seek the covert in their fear.

But soon the cabin's lazy smoke
   I see above the orchard curl;
And, hark! what sound the silence broke?
   The jocund laugh of boy and girl!

Around and round, in merry rout,
   I see them go, as though to play
Were all of life, and care and doubt
   Could never cloud their summer day.

The oriole her pendent nest
   Is hanging from the willow bough;
The lark with joy distends his breast,
   And warbles to the lowing cow.

Thus Nature everywhere repeats
   The beauty and the love she owns;
From hill to sea her rhythmic beat
   Is heard in many blending tones.

And Art, her handmaid, catches up
   The glory of each sound and sight,
To pour them from her magic cup,
   A draught to steep us in delight.

A WILD NOSEGAY.

Sweet-scented messengers from landscape green,
Thy presence is a blessing in my cot,
A still memento of each sunny spot,
Or shaded, where my wandering feet have been
In search of thee. The winding, wet ravine,
Luxuriant with golden flowers; the grot
Beneath the live-oak, where small blossoms dot
The mossy rock, and humming-birds are seen
To flash and quiver through the tremulous leaves
Of snowy buckeye; and the mountain steep
Or wooded summit, where sad zephyr grieves
Forever through the branches of the pine;—
All helped to form thee, and thou still dost keep
Their charms before me, which I blend with thine.

MOUNTAIN, LAKE, AND VALLEY.

In attempting a general introductory view of the scenography of California, we shall be aided by an outline of its topography. The materials for this are to be found mainly in the preliminary report upon the geology of California by Prof. J. D. Whitney. Before the great work conducted by him was begun, hardly fourteen years ago, there was little exact knowledge of the physical structure of the Golden State. Its broadest features were known in a general way; but some of the most remarkable regions were unexplored, and a mass of interesting details had been only casually observed, if at all. An adventurous and daring people, engaged in the stimulating search for gold, had revealed the secrets of many places which would else be blank spaces on the maps; but the area of a territory larger than New England, New York, and Pennsylvania combined, and embracing two mountain chains surpassing in some respects the Alps and Appalachians, could not be thoroughly explored and accurately described without concerted effort to that end. When that effort — temporarily abandoned through a freak of ignorant legislation — shall be resumed and completed, we shall have, in a series of valuable reports even now far advanced, ample material for special studies. In the mean time, even such a mere sketch as we shall offer of the valley and lake system of California may prove interesting to the general reader. The topography of California is characterized by a grand simplicity. Two mountain-chains — the Coast Range and the Sierra Nevada — outline the form of the State; the one extending along the Pacific shore, on its western side; the other, along its eastern border, overlooking the great basin of the middle continent; and both interlocking north and south, inclosing the broad, level valleys of the Sacramento and San Joaquin. The axial lines of these chains have a northwesterly and southeasterly course. They are clearly distinguished between the thirty-fifth and fortieth parallels — the valleys named, which have a length of nearly three hundred and fifty miles, and a breadth of from forty to eighty miles, separating the two systems completely. North and south of the limits named, the Coast Range and the Sierra Nevada are topographically one, distinguishable only by geological differences; the former having been uplifted since the cretaceous deposition, and the latter before that epoch. The Coast Range is inferior in altitude, averaging only from two thousand to six thousand feet above the sea, and having few prominent peaks. It extends the whole length of the State, say seven hundred miles, and has an aggregate width of forty miles; but it is broken into numerous minor ridges, marked by striking local differences, and separated by an extensive series of long, narrow valleys, which are usually well watered, level, fertile, and lovely. The Sierra Nevada has an altitude of from four thousand to twelve thousand feet, and an average width of eighty or one hundred miles. It rises from the central valley in solid majesty, reaching by a gradual slope its double crests, which culminate in a nearly straight line of peaks extending a distance of five hundred miles. There is no peak in the Coast Range which rises above eight thousand feet. The Sierra Nevada has a hundred peaks which rise about thirteen thousand feet, and at least one which soars fifteen thousand feet. Where the two ranges join at the north (latitude forty degrees, thirty-five minutes), Mount Shasta, which may be taken as a point of connection, attains an elevation of fourteen thousand four hundred and forty feet. Its snowy summits can be seen from great distances in Oregon, California, and Nevada, and is nearly twice the height of any other mountain in its vicinity. As the Sierra Nevada extends southward from this point, it gradually increases its general altitude. For three hundred miles the passes range from four thousand to eight thousand feet above the sea, and the peaks from one thousand to two thousand feet higher. But from latitude thirty-eight degrees, for a distance of two hundred miles along the summit, there is no pass known lower than eleven thousand feet, and within that distance all the chief peaks have an elevation of thirteen thousand feet.

The summits of the Coast Range are only occasionally whitened with snow in the winter. Those of the Sierra Nevada are covered with it every winter to a great depth, and on some of them it never melts. The Coast Range rises with tolerable abruptness facing the sea, its inner line of ridges sloping gradually to the central valley. The Sierra Nevada has a gradual ascent on its western side, but an abrupt one on its eastern, the latter being only half as long as the former, since it meets the elevated plateau of Nevada or Utah, four thousand to five thousand feet above the sea. The Coast Range is broken near its centre, at the Golden Gate, where the Bay of San Francisco receives and discharges the waters of the Sacramento and its tributaries, forming the river system of the whole northern interior; and those of the San Joaquin, forming the river system of the southern interior as far as the Alpine region of the Sierra. The Sierra Nevada is unbroken in its whole length, although the table-lands and depressions at its northern and southern extremities are nearly on the level of the plateau to the eastward, and offer the easiest wagon and railroad approaches from that side. The most striking feature of the vegetation of the Coast Range is its majestic groves of redwood, which flourish only in the foggy regions north of San Luis Obispo, and in connection with a soil overlying a metamorphic sandstone. The inner ridges of the Coast Range are frequently bare, or covered chiefly with varieties of oak, interspersed with the madroña, remarkable for its smooth, bronzed trunk, its curling bark, and its waxen leaves. When not tree-clad, these inner ridges, to a height of from five hundred to twenty-five hundred feet, are covered with wild oats, and suggest the idea of immense harvest-fields that have been thrust up by volcanic energy, and left standing high in the blue air. As the state geologist says:[1]

[1] *Yosemite Guide-book*, p. 35.

FOOT-HILLS OF THE COAST RANGE

" What gives its peculiar character to the Coast Range scenery is, the delicate and beautiful carving of their masses by the aqueous erosion of the soft material of which they are composed, and which is made conspicuous by the general absence of forest and shrubby vegetation, except in the cañons and along the crest of the ranges. The bareness of the slopes gives full play to the effects of light and shade caused by the varying and intricate contour of the surface. In the early spring these slopes are of the most vivid green, the awakening to life of the vegetation of this region beginning just when the hills and valleys of the Eastern States are most deeply covered by snow. Spring here, in fact, commences with the end of summer; winter there is none. Summer, blazing summer, tempered by the ocean fogs and ocean breezes, is followed by a long and delightful six months' spring, which in its turn passes almost instantaneously away, at the approach of another summer. As soon as the dry season sets in, the herbage withers under the sun's rays, except in the deep cañons; the surface becomes first of a pale green, then of a light straw yellow, and finally, of a rich russet-brown color, against which the dark green foliage of the oaks and pines, unchanging during the summer, is deeply contrasted." The most striking feature of the vegetation of the Sierra Nevada is its magnificent growth of pines, comprising several species which attain a height of from one hundred and fifty to three hundred feet, and the famous groves of *Sequoia gigantea*, which equal in height, if not in age, the pyramids of Egypt. The prominent lithological feature of the Coast Range is the prevalence of metamorphic cretaceous rocks. The lithological structure of the Sierra Nevada is more primitive, granite being the prominent feature, underlying a greater part of its extensive beds of auriferous gravel, and giving an air of gray desolation to its naked summits, which bear the marks of ancient glaciers. The Sierra Nevada is also distinguished for the evidences it presents of the tremendous forces that raised it at three successive epochs above the sea. A hundred volcanoes have blazed along its crest, and have covered with lava an area of not less than twenty thousand square miles, not uniformly level or sloping, but seamed with cañons hundreds or thousands of feet deep, through which flow the living streams of the Sierra. Sometimes this lava overlies, and at others underlies, the deposits of gold-bearing gravel wrought by the miner. Sometimes the eruptive rocks, contemporaneous with its flow, rise in picturesque crags that rival in height the summits of the older granite.

This glance at the mountain frame-work of California is necessary to an understanding of its lake and valley system. The chief feature of this system is the central valley of the Sacramento and San Joaquin, supplemented at the south by the valleys of the Tulare and Kern. These valleys form a basin about four hundred miles long by fifty or sixty miles wide, which was anciently the site of lacustrine or marine waters. In its northern portion rises abruptly from the level plain a singular local mountain ridge, known as Sutter's Buttes, which is an object of beauty in the landscape views of that region, and seems, in the flooded seasons, like an island in the main. North of the Buttes the valley gently swells to meet the foot-hills of the blending Sierra and Coast Range; and these uplands consist of a red and gravelly soil, whereas the general surface of the valley southward is a rich, deep loam, which has frequently been known to yield from sixty to seventy bushels of wheat to the acre. The climate of this fertile basin is very warm in summer, and favorable to the out-door growth of roses and strawberries in winter. It is timbered at intervals with open parks of oaks, which become more numerous near the foot-hills on either side, and there mix with inferior coniferæ and minor vegetable forms, including the characteristic manzanita, buckeye, and laurel. The principal rivers are fringed with sycamore, oak, cottonwood, willow, alder, and white maple. Sweet-briers bloom close to the streams, and, where the timber has not been cut away, the wild grape-vine still hangs its graceful curtains, through which the boatman catches glimpses of beautiful woodland or valley scenes, and a far background of hazy mountains. Immense tracts are annually covered with a luxuriant growth of wild oats, which, alternately green or gold, according to the season, rolls its surface in rippling light and shade under every breeze. The moist bottoms yield heavy crops of grass. In the spring, the whole surface of these valleys, where not cultivated, is thickly covered with wild flowers of every color; and the scene of this gay parterre, broken with seas of verdant grain, and bounded by walls of blue or purple mountains, whose peaks are capped with snow, is quite entrancing. These charming plains were the favorite resort of the aborigines, who found in the streams that drain them plenty of salmon, sturgeon, and lesser fish, and all over their extent herds of antelope and elk, and myriads of ducks and geese, besides quail, doves, hares, rabbits, and squirrels. The grizzly would sometimes come from the hills to eat fish and berries; but he was game beyond the skill of the simple savages who once enjoyed the central valley alone. Into the rivers discharge the numerous channels which cut the western slope of the Sierra, receiving the heavy rains that wash its flanks, and the melting of the deep snows upon its summit, and almost annually the accumulated torrents overflow portions of the level land.

There are no lakes in the central valley, except in its lower extremity, where Tulare Lake, thirty-three miles long by twenty-two wide, surrounded by a broad area of reedy marshes, forms the mysterious sink for all the streams coursing down the western slope of the southern Sierra. The general features of the valleys in Fresno, Tulare, and Kern counties, are not essentially different from those of the Sacramento and San Joaquin, which they supplement. The chief point of difference is their hydrography. There are considerable tracts of marsh land in the larger valleys named, but they are formed by the rivers and estuaries of the central bay; while those of the lower valleys are an adjunct of the lakes, about which they comprise an area of fully two hundred and fifty square miles.

Most of the streams of the central valley flow from the Sierra Nevada. A dozen principal branches of the Sacramento and San Joaquin rivers, and the rivers that sink in the Tulare Lake, are fed along a distance of four hundred miles, from Shasta to Tejon, by several hundred tributaries which rise in that great chain. In the same distance a few score of creeks flow eastward from the inner ridges of the Coast Range, to the central basin, and some of these are dry in the summer. The small rivers of the Coast Range flow through the intervales, emptying either into the ocean at right angles to the trend of the coast, or following the valleys parallel with the trend till they reach some of the bays that make inland.

The valleys in the Coast Range are numerous and dissimilar, though possessing some marked characteristics in common. Those of one class lie open to the sea, and are usually narrow, with a trend nearly east and west, or following that of the coast. Most of them are found south of the Bay of San Francisco, itself skirted by a series of valleys which slope from the base of the Mount Diablo range. The largest of the coast valleys is the Salinas, in the Santa Cruz and Monterey district. It is about ninety miles long by eight to fourteen miles wide, mostly arable, and yielding heavy crops of wild oats and clover. Although the open coast valleys are subject to the winds and fogs, they possess a fine climate, and are cultivated to the very margin of the sea. It is a beautiful sight to behold their grassy margins skirting the crescent lines of small bays, or their wide fields of yellow grain contrasting with the blue line of the ocean, while behind rise the rumpled velvet of bare hills, tawny or verdant, with the season, and the farther crests of cloud-girt summits bristling with redwood forests that keep moist in the salty air. Perhaps the most picturesque valley that opens to the sea, though it meets the ocean only at its extremity, is Russian River Valley, north of San Francisco. It is long and narrow, has a generally level but sometimes rolling surface, is traversed by a clear stream, and bounded on either hand by ridges, which have a great variety of form. Its groves of oak, its picturesque knolls, its vistas of conical peaks, its winding stream, alternately placid and rapid, its luxuriant carpet of grass, grain, and flowers, have long made it a favorite sketching resort for artists. The valleys of Mendocino, still farther north, are smaller, but possess scenery of more grandeur, and are remarkable for the number of streams that flow through them to the sea. Humboldt County, also, has some picturesque valleys, that look out upon the sea, or line the bay which bears the name of the author of "Cosmos."

The inner series of Coast Range valleys is the most extensive. While the outer valleys are generally separated by abrupt and treeless ridges, those inland are divided by gentler elevations, which are covered by trees or clad with grass and wild oats. The inner valleys, again, lie parallel to the trend of the coast. They are commonly oblong, nearly level, or rolling like the Western prairies, extremely fertile, and have a climate more sheltered from the sea-wind and fog.

Among the most celebrated of these are the Sonoma, Napa, Santa Rosa, Suisun, Vaca, Berreyesa, and Clear Lake, north of the Bay of San Francisco, and some of them communicating with it; and the Alameda, Santa Clara, Amador, Pajaro, and San Juan, to the east or south of the bay. An enumeration of all the coast valleys distinctively known, would be a tedious task. They are the favorite nestling places of our population, as they were the favorite sites of the Mission Fathers, and offer examples of the most elaborate cultivation, the most contentment, and the greatest thrift. Seldom more than three or four miles wide, often not more than one, they are in length from five to fifty. Their gently rolling surfaces rise into mound-like hills on either side, — the best soil for the wine-grape, — which in turn are flanked by ridges or peaks from five hundred to perhaps three thousand feet high. The creeks with their dark green belts of timber, often live-oak, wind through continuous harvest-fields. Many of the farm-houses are prettily built on knolls that command a good view. Nothing can be finer than the aspect of many of these valleys, when the lush verdure of the early spring is prodigally gemmed with wild blossoms of the most brilliant colors, or when the rich gold of their summer fields, islanded with the clumps of evergreen oaks, is contrasted with the purple or blue mountain, and the sky at morning or evening brightens or fades through tints of amber and amethyst. Sometimes the splendor of the setting sun seems to penetrate the dark substance of the solid hills, and give them a transparent glow, as if they yet burned with the heat of their thrusting up. As light comes in the spring or summer, the trees are vocal with linnets, while larks sing in the fields, and chanticleer sounds his horn. As day goes, it is pleasant to hear the birds calling to repose, the wild doves cooing, the quails fondly signaling their mates, the owl adding his solemn note to the vespers of the feathered tribe. One thinks of the day when a native generation will love these mountain-walled valleys, with their wealth of varied scenery and resources, as ardently as the "pioneers" loved the home-spots which they left in the East or in Europe. Poetry and song and romance will come at last to link the spells of imagination and fancy to those of memory and affection, and "home" will exist here as, in the fond old meaning of the most characteristic English word, it exists now for so few.

The coast valleys are too near the level of the ocean, and the mountains surrounding them are too broken, to contain many lakes. Few are known which deserve description; but one of these, in Lake County, about eighty miles north of San Francisco, is one of the most remarkable and lovely in the State. It is called Clear Lake, in spite of the fact that, owing to its shallowness and the easy disturbance of its muddy bottom by winds, it is scarcely ever clear. Seen from an elevation, however, as it reflects the color of a seldom-clouded sky, it loses nothing by comparison with purer sheets of water. It is a pity that its Indian name of Lup Yomi, whatever its meaning, could not be substituted for its present commonplace title. Clear Lake lies in a valley between two ridges of the Coast Range, thirty-six miles from the ocean, and has a length of twenty-five miles by a width of from two to ten miles. Its elevation above the sea is about fifteen hundred feet. The region surrounding it is ruggedly mountainous, and embraces an ancient volcanic centre. St. Helena, at the head of Napa Valley, to the south, and the highest peak between San Francisco and the lake, is an extinct volcano, and the evidences of its former activity are abundant for many miles in every direction. Midway between this peak and the lake are the famous geysers, and mineral springs and deposits are frequent throughout the whole region. For several miles the road approaching the valley from the direction of Napa passes over a mountain largely of obsidian. The cuttings through this material reveal it boldly; the undisturbed surface is covered with boulders and cobbles of it, and in the roadway it is ground into pebbles and sand. A deluded person who was convinced that it was glass and could be readily manufactured, once sank considerable money in a vain attempt to convert it into bottles. On the western shore of the lake, not far from the base of this obsidian mountain, is Borax Lake, a small and shallow pond remarkable for the large percentage of borax contained in its waters and muddy bed. Extensive deposits of sulphur are also found in the vicinity. The inclosing ridges are peculiarly rugged, and the conical peaks numerous. One of these, called Uncle Sam, rises abruptly from the edge of the water to a height of twenty-five hundred feet, dividing the lake into two parts, Upper and Lower. Near the upper end of the lake Mount Ripley attains an elevation of three thousand feet, and farther off rises Mount St. John, nearly four thousand feet. Still higher peaks, on the northeastern side, bearing aboriginal names, are often covered with snow, and at such times the traveler descending to the lake from the west, and seeing these white peaks beyond the blue expanse and green margin of meadow and grove, is reminded of Switzerland and the Alps. Where not volcanic the rocks are cretaceous, and abound in fossils. Ridges of serpentine occur, which are richly charged with quicksilver. Formerly the lake must have filled the whole valley, covering even the low ranges of adjacent sand-hills, which afford every mark of recent denudation, and are eroded into mound-like forms of striking regularity. Upper Lake is nine miles wide; Lower Lake is much narrower, but contains several pretty rounded islets, bearing a golden harvest of wild oats, shaded by orchard-like white oaks, and still partly occupied by Indians, who live chiefly on the trout, pike, and black fish which they catch in the water, and the ducks, geese, and other wild fowl which tenant its reedy shores. Deer and bear abound in the well-wooded mountains. Several streams put into the lake, and one flows from its lower extremity, emptying into Cache Creek, a tributary of the Sacramento. Northeast of Uncle Sam lies a fine valley, the seat of a thrifty community. Its rich loam bears a noble growth of ancient and mighty oaks, among which nestle sundry villages.

In the northern part of California, where the Coast Range and Sierra Nevada interlock, the system of valleys is confused and difficult to describe. Yet it may be said that they preserve the oblong form and level surface which characterizes the entire family of Pacific valleys. The upper part of the Coast Range proper, extending to and including the Humboldt Bay country, comprises a noble series of pastoral and agricultural valleys, watered by streams rich in salmon and flanked by mountains which are covered with forests of the stately redwood. Some of these valleys were the scenes of conflicts with Indians for many years, and owe their sparseness of population partly to this cause and partly to their isolation. In rugged Trinity County there are only a few small valleys along the water-courses. In Klamath County the largest valley is Hoopa, thirty miles long and two wide, at the junction of Trinity and Klamath rivers. Del Norte has a number of small, fertile valleys. Siskiyou has the largest valleys of any of the northern counties. They seem to be intimately connected with the plateau east of the Sierra, and to have some of its characteristics. Scott Valley, forty miles long by seven wide, lies between the Trinity and Salmon ranges, which are six thousand feet high, the valley itself having an altitude of three thousand feet, and possessing a climate more like that of some of the Northern States than the lower valleys of California. Surprise Valley, in the extreme northeastern part of the State and overlying the Nevada boundary, is sixty miles long by fifteen wide. It has an elevation even greater than Scott Valley, but it is as fertile as it is lovely. Its ample surface is finely watered, and covered with a rank growth of native clover and grass, on which feed immense flocks of wild geese and brant in their season. On its east side are three beautiful lakes, which extend nearly its whole length, and cover almost half its surface. They contain no fish, but are the resort of great quantities of ducks, geese, cranes, pelicans, and other wild fowl. They receive a number of small streams, but have no outlet. Shasta and Elk valleys are lava plains, three thousand to three thousand seven hundred feet above the sea. They are remarkable only for the fine views they command of

Mount Shasta, and the former for the numerous small volcanic cones that dot its surface. The Shasta region is only the southern extremity of that vast volcanic territory which includes the famous Modoc lava beds, and which, extending into Idaho and eastern Oregon, including the country drained by the Columbia and Snake rivers, embraces an area of nearly three hundred thousand square miles, which is overlaid with lava hundreds and thousands of feet thick, covering ancient forests and mammoth skeletons.

Siskiyou County contains a number of large lakes besides those in Surprise Valley. Its total lake surface is equal to half a million acres. Klamath Lake, the source of Klamath River, lies partly in this county and partly in Oregon. Eastward from it, lying wholly in Siskiyou, are Goose, Rhett, and Wright lakes, which are the sources of several rivers traversing the northern counties of California, including the Trinity, Salmon, and Pitt. The last named river debouches from Goose Lake, which is thirty miles long and sixty wide, and is surrounded by a fertile valley of thirty or forty thousand acres.

Leaving Siskiyou, whose vales and plateaus, sterile plains of lava, and wide but shallow sheets of water, have an elevation of from three thousand to four thousand feet above the sea, we reach the simple topography of the Sierra, with its regular ridges leading to lofty peaks, and divided by profound cañons. Here one would scarcely expect to find valleys; yet there are hundreds of small valleys in the lofty chain, many of which are inhabited and cultivated. One series of valleys, and these are the smallest, lie along the watercourses on the western flank of the Sierra, at right angles to the trend of the range, and frequently forming the passes by which it is crossed. Another series lie between the double crests of the summit, parallel to the trend of the chain. The valleys on the two flanks form convenient roadways, and were followed by the first emigrants to California. The famous Beckworth, Henness, and Truckee routes across the Sierra Nevada all lie through a succession of such small intervales, reaching on either side of the Sierra to an open and level pass. The Pacific Railroad crosses the Sierra partly by the aid of these natural road-beds, following the course of the Truckee down

Yosemite Falls, from Glacier Point.

the eastern slope. The most remarkable of these transverse valleys partake of the nature of gorges.

One of them, the Yosemite, has a world-wide celebrity. The valley itself is an almost level area, about eight miles long and from half a mile to a mile in width. Its elevation above the sea is four thousand feet, and the cliffs and domes about it are from seven thousand to nine thousand feet above the sea, with an altitude above the valley of from three to five thousand feet. Over these vertical walls of bare granite tumble the Merced River and its forks. Most of the cañons and valleys of the Sierra have resulted from denudation, and some have been partly shaped and marked by glaciers; but Professor Whitney thinks that this mighty chasm has been roughly hewn in its present form by the same kind of forces which have raised the crest of the Sierra and moulded the surface into something like their present shape. He conceives the domes were formed by the process of upheaval itself, and says that the half dome was split asunder in the middle, the lost half having gone down in what may truly have been said to have been "the wreck of matter and the crash of worlds." John Muir, who combines the feeling of a poet with the patient observation of a scientist, and who spent several years of research in this part of the Sierra, contends, on the contrary, that glacial action was the main force which sculptured this wonderful fane of nature. Another gorge, which is inferior only to the Yosemite, is found at the sources of the Tuolumne River, still farther in the heart of the Sierra. Its vertical cliffs would be unique in the mountain scenery of the world, were Yosemite unknown. It is here that the tourist approaches the Alpine region of California. The summit of the pass leading into Tuolumne Valley is nine thousand and seventy feet above the sea, and the descent to the river is only about five hundred feet. Tenaya Valley, between Yosemite and Tuolumne, contains a beautiful lake by the same name, a mile long and half a mile wide. A high ridge near this lake commands a fine view of Cathedral Peak, which Professor Whitney describes as a lofty ridge of rock cut down squarely for more than one thousand feet on all sides, and with a cluster of pinnacles at one end, rising several hundred feet above the rest of the mass. It is at least two thousand five hundred feet above the surrounding plateau and eleven thousand feet above the sea. At the head of Lake Tenaya rises a conical knob of bare granite, eight hundred feet high, its sides finely polished and grooved by former glaciers. The upper Tuolumne drains a richly turfed valley half a mile or a mile wide, and fifteen miles long, and containing some noted soda springs. The valley has an elevation of from eight thousand six hundred to nine thousand eight hundred feet. In this vicinity are the most remarkable evidences of the former glacial system of California. The whole region rapidly rises till it meets the dominating peaks of the King's River country.

The highest of the transverse valleys is Mono Pass, which is ten thousand seven hundred and sixty feet above the sea; and the most elevated pass used by travelers is the Union. In a cañon at the eastern side of this pass are several small lakes, not less than seven thousand feet above the sea, which are produced, like many of the lakes of the high Sierra, by the damming of the gorge by the terminal moraines left by the retreating glaciers. Mount Dana is the culminating point of the Sierra in the region of the upper Tuolumne. It has an altitude of thirteen thousand two hundred and twenty-seven feet. To the east of it, only six miles, but nearly seven thousand feet below, lies Mono Lake, a body of water fourteen miles long from east to west, and nine miles wide, highly charged with mineral salts, void of all life except the countless larvæ of a small fly, sluggish and dreary in appearance, and surrounded by strong tokens of smouldering volcanic agencies, among which is a cluster of truncated cones.

Below the region of the high Sierra in Southern California, the valleys or table-lands connect with the Nevada plateau, or Great Basin, and are mainly of the same character — arid, alkaline, and barren. The streams flowing east or west are bordered by narrow strips of level land, supporting tuft grasses, willows, and cottonwoods, but offering little inducement for settlement. There are numerous salt lakes and ponds. The largest of these is Owen's Lake, twenty-two miles long and eight wide. In the same region, lying partly in San Bernardino and partly in Inyo counties, between Owen's Lake and the Nevada line, is Death Valley. This remarkable depression is the lower sink of the Amargosa River, and, although situated in the high Sierra, it is actually one hundred and fifty feet below the level of the sea. The soil is a thick bed of salt, and, doubtless, the depression was formerly occupied by a lake. All the salt lakes of the region we have described have marked in terraces their former larger dimensions, and are evidently in process of gradual extinction. This portion of the Sierra has been frequently disturbed by violent earthquakes within a few years past. Some of these shocks have been followed by a rise in the waters of Owen's Lake, which continued until it had overflowed thousands of acres, and then suddenly abated, the lake resuming its usual size.

While the valleys and lakes of the Tuolumne and

King's River region present altogether the strongest and grandest features, those between this region and the sources of Feather River northward are the most pleasant. All the rivers in this stretch of country flow partly through small valleys; but the larger valleys are those of the summit, lying between the crests of the Sierra, or on its flank, from three thouand to seven thousand feet above sea level, while the ridges that inclose them on the east and west rise from one thousand five hundred to three thousand feet higher. The largest of these valleys lie at the sources of the Feather River, in Plumas and Lassen counties, connecting with easy approaches from the Nevada plateau, and offering low and comparatively snowless passes for winter transit of the mountain. Honey Lake Valley, in Lassen, contains about twenty thousand acres of meadow and arable land, is one of the lowest in altitude, and possesses a mild winter climate. The lake from which it is named is twelve by five miles in dimensions, of irregular form, and constantly decreasing size. It is really an independent basin, lying east of the Sierra crests, and receives the water of two rivers. The valley is sixty miles long by fifteen to twenty wide. It is named from the quantity of honey-like liquid deposited plentifully on the grass and shrubs by a species of bee peculiar to dry and barren countries. Eagle Valley contains a shallow and irregular lake, about twelve miles long by eight wide. Long Valley, in the southern part of the county, is about forty miles long by two or three wide, quite level, and notable for its superior pasturage. Southward of this valley, the summit valleys decrease in size with increase of altitude. While the Lassen and Plumas valleys are only from three thousand to four thousand feet above the sea, those in Sierra, Nevada, Placer, El Dorado, and other counties to the southward, are from five thousand to seven thousand feet high. A third small lake in Lassen, called Summit Lake, has an altitude of five thousand eight hundred feet, with a little strip of level land. Plumas contains nearly a score of valleys that are fertile, sheltered, and populous, lying on the upper tributaries of Feather River, and embracing an aggregate of nearly two hundred and fifty thousand acres of good land. The snows are light in these valleys.

All the lesser summit valleys have characteristics in common, varying chiefly as to size and altitude. They are usually long and narrow, covered with a luxuriant growth of natural grasses, watered by small willow-fringed streams that flow either west or east, gemmed by small lakes, and framed by more or less rugged ridges, bearing thick forests of pine and fir to near their summits, which are bare crags of gray granite, covered for a great part of the year with snow. The discovery of silver in Nevada, in 1859, and the subsequent settlement of that State, brought these valleys into notice and use. Before that event, they were mostly resorted to by drovers for summer pasturage, cattle being driven thither from the parched plains of California in summer, and brought back on the approach of winter. At a later day their grasses were cut for hay, to be sold in Nevada, and to way-travelers. Many of them lay directly in the numerous routes leading from California to the silver regions, and began to be appropriated by settlers for ranching and lumber purposes. Finally the building of the Pacific Railroad has given many of them special value, and some of them are becoming places of great resort for summer tourists, invalids, and artists. It is certain that most of them will soon be occupied by permanent communities, and that the Sierra Nevada will ultimately contribute a stream of hardy life to counteract the enervating effect of extreme heat in the lowlands of California. Their summer climate is delightfully temperate and bracing; their winter climate cold, but seldom extremely so. Those which are most sheltered, and not too high, produce whatever will mature in New England. In others, the growing season is too short for much effective cultivation; but lumbering, mining, and quarrying will furnish employment for considerable settlements, and markets for the products of more favorable spots.

The most attractive feature of the lesser summit valleys is their multitude of clear, fresh lakes, stocked

with the finest trout, surrounded by magnificent groves of pine and fir, reflecting snowy peaks, and beautiful with all the colors of changing day and evening. Concerning this charming feature, less has been reported than of any other. A standard authority on the physical features of California has even made the broad assertion that the Sierra Nevada contains very few lakes. This mistake was natural; for, aside from the singular salt or alkali lakes in the volcanic regions of the Sierra, north and south, together with the few large fresh-water lakes already enumerated in this article, the lakes of the Sierra have not been mapped or described. On no popular chart of this range are more than twenty or thirty lakes indicated, whereas the existence of at least two hundred, in a distance of four hundred miles, from Siskiyou to Kern, can be positively vouched for; and this number is probably within the truth, as it will be developed by future explorations. These lakes are the sources of the numerous rivers that have eroded the deep cañons of the western slope, and of the few which flow eastward.

They are the reservoirs of the melting snows — the sources of summer supply for hundreds of miles of mining ditches. Some are sunk deep in rocky chasms, without level or meadow land surrounding them. Others have been formed by glacial moraines damming up the gorges that would else have been only the channel of streams. Nearly all have been larger and deeper than now. Some are no larger than the petty tarns of the English hills; while others would float a navy, and can mimic the commotion of the sea.

Sierra County contains twenty or more small lakes, situated in the depressions of the summit, generally circular in form, from half a mile to a mile across, and varying in depth from a few feet to ten or twenty fathoms. The largest, Gold Lake, about four miles long by two wide, is famous as the scene of falsely reported deposits of lump gold, which, in 1849-50, attracted and disappointed a multitude of miners. Nevada County, next adjoining Sierra on the south, is still richer in lakes, containing at least thirty. Four of these are notable as the sources of supply for one of the most extensive mining canals in the State, — that of the Eureka Lake and Yuba Canal Company. The trunk canal of this company is sixty-five miles long. Its principal supply reservoir is Eureka Lake. This originally had an area of only one square mile, but an artificial dam of granite across the outlet, one hundred and twenty feet long at the base, two hundred and fifty feet long at the top, and seventy feet deep, has doubled the surface of the lake, and given it an average depth of sixty-five feet. Lake Faucherie, with a wooden dam thirty feet high, floods two hundred acres. Two smaller lakes with these feed a canal eight feet wide by three and a half feet deep, and furnish water for some of the heaviest deep-gravel mining in the State. The South Yuba Canal Company has utilized five lakes in another part of Nevada County. One of these, Meadow Lake, is enlarged by a solid masonry dam, which is forty-two feet high and eleven hundred and fifty feet long, and makes, when full, a sheet about two miles long by half a mile wide, with a depth varying, according to the season, from ten to thirty fathoms. Seven miles in a southeasterly direction are White Rock, Devil's Peak, and two smaller lakes which, jointly, equal the capacity of Meadow Lake. Devil's Peak Lake lies close to the Pacific Railroad. These reservoirs are drawn into the channel of the South Yuba, when that stream runs low in the summer, and thence pass through fifty miles of ditching.

The works of the two companies named cost an aggregate of several million dollars. When they cease to be wanted for mining purposes, they will serve to irrigate countless gardens and vineyards on the lower slopes of the Sierra. Meadow Lake gives a name to a large township which is remarkable for being one of the highest mining localities in California, as for the great size and number of its gold and silver ledges. The general altitude of the district is from seven thousand to eight thousand feet, and contains about twenty lakes. Snow fell there in the winter of 1866–67 to the depth of twenty-five feet, yet many daring people remained and mined through the season, and several towns are growing up. Within the district are Crystal and Donner lakes — the former one of the most picturesque resorts in the Sierra; the latter having a beauty of another kind, and being remarkable as the scene of a painful tragedy in the early settlement of the State. Donner Lake is three miles long by one wide. It lies in sight from the eastern end of the summit tunnel of the Central Pacific Railroad, one thousand five hundred feet below that point and five thousand five hundred feet above the sea. A small stream pours from it into the Truckee River, only three miles eastward, watering a narrow valley. Here, late in October, 1846, a party of eighty overland immigrants, under the lead of Captain Donner, and including over thirty women and children, were overtaken by a snow-storm, which prevented them from proceeding. They suffered terribly in their winter camp, or while wandering blindly searching an outlet, until found by relief parties from the western side of the mountains, in February. In the sequel, thirty-seven perished from exposure and hunger, and some of the party were only sustained by the last dreadful resort of starving humanity. The locomotive now almost hourly passes the scene of this tragedy, awaking clanging echoes among the dizzy cliffs of bare granite through which its way is cut. Hundreds of people live in or about the valley the whole year; and hard by thirty saw-mills are busy thinning out the noble forests that deck the steep slopes on every side.

A congeries of small lakes is found to the southward of the Pacific Railroad where it crosses the summit, each of which has its peculiar charms, and its special friends among the tourists, who begin to seek these sylvan sheets through the warm season. They lie from six thousand to seven thousand feet above the level of the sea, where the snow falls commonly ten feet deep, and stays from November or December until July, with lingering patches sometimes on the peaks above until the next winter. Some of these lakes are appropriated for ice supplies to the lower country. Rude hotels have been erected near a few, to accommodate the visitors who go there to fish, sail, sketch, and recuperate.

All the lakes of Sierra and Nevada counties, except one or two, — like Donner, which lies on the eastern side of the summit, or Truckee, which is just over the line of gradual eastern descent in the Henness Pass, and feeds Little Truckee River, — are sources of the numerous tributary streams that feed and form the Yuba River, or the northern forks of the American. Another congeries of small lakes in Placer and El Dorado counties feed the larger forks of the American and Cosumnes, and supply an extensive system of mining canals. The South Fork Canal, one of the largest of these works, having a length of one hundred and forty-two and a half miles, is partly supplied from Silver, Red, and Willow lakes, which store up together nearly three hundred and fifty million cubic feet of water. Some of this goes to irrigate the vineyards for which the high, red hills of El Dorado are becoming celebrated. Through the whole middle tier of mining counties, from Siskiyou to Mariposa, the summit lakes are more or less drawn upon to fill artificial channels, and aid in the extraction of gold and the cultivation of the soil. Their names make a long list, and suggest their picturesque qualities, — as Silver, Crystal, Cascade, Emerald, Grass, Fallen Leaf, Tule, Willow, Mirror, Alder, Palisade, etc. Many are named from the peaks that overlook them, from the wild animals or birds that frequent them, from the circumstances of their discovery, or from the persons who first took up abodes near them. The most extensive and celebrated of the whole group is Lake Tahoe, in El Dorado County, only fifteen miles southwardly from Donner Lake and the line of the Central Pacific Railroad. Its elevation above the sea, exceeding six thousand feet; its great depth, reaching a maximum of more than one thousand five hundred feet; its exquisite purity and beauty of color; the grandeur of its snowy mountain walls; its fine beaches and shore groves of pine, — make it the most picturesque and attractive of all the California lakes. Profound as it is, it is wonderfully transparent, and the sensation upon floating over and gazing into its still bosom, where the granite boulders can be seen far, far below, and large trout dart swiftly, incapable of concealment, is almost akin to that one might feel in a balloon above the earth. The color of the water changes with its depth, from a light, bluish green, near the shore, to a darker green, farther out, and finally to a blue so deep that artists hardly dare put it on canvas. When the lake is still, it is one of the loveliest sights conceivable, flashing silvery in the sun, or mocking all the colors of the sky, while the sound of its soft beating on the beach is like the music of the sea-shell. When the wind angers its surface, its waves are dangerous to buffet. The sail that would float over its still face like a cloud is then driven like fate, and is lucky to escape destruction. Sometimes the dense ranks of tall pines, firs, and cedars extend to the shore and are reflected in the placid sheet. There is always some new beauty to see, and one scarcely knows which is most delightful, — to float over the deep blue element that kisses his bark, or to wander along the sandy beach and through the surrounding woods, thinking of the power that reared this noble range and gemmed its deep gorges with such scenes of witchery.

A MEMORY OF THE SIERRA.

My heart is in the mountains, where
They stand afar in purple air.
Up to their peaks and snowy founts
In happy dreams my spirit mounts.
Their ridges stretch unto the plain,
Like arms, to draw me up again;
The plain itself a pathway is
To lead me to remembered bliss.

I hear the brown larks tune their lay,
And little linnets, brown as they,
Fill up the intervals with sweet
Enticement to their green retreat.
I hear the wild dove's note forlorn,
The piping quail beneath the thorn,
The squirrel's busy chip and stir,
The grouse's sudden heavy whir,

The cawing of the white-winged crow,
And chatter of the jays below.

I stand within the cloistered shade
By columned fir and cedar made,
And up the minster-mocking pine
I gaze along the plummet line
Of mighty trunks, whose leafy tops
Distill a spray of diamond drops,
Whene'er the sunlight chances through
Their high mosaic of green and blue.
I hear a sound that seems to be
An inland murmur of the sea,
Yet know it is the tuneful moan
Of wind-touched forest harps alone.

I wander to the dizzy steep
That plunges into cañon deep,
And where the obscuring hazes hint
The amethyst and violet's tint.
I see along the cloudless sky
My dear-loved peaks, serene and high —
So cold at morn, but warmly bright
With flushes of the evening light.

A MEMORY OF THE SIERRA.

The very eagles hate to leave
These heights sublime, but fondly cleave
In circling flights about the crests
Where they have built their lonely nests.

Perched on these crags, the world below
Melts in the hazy summer's glow;
Hid are its gloomy sounds and sights
From all who reach these templed heights,
Where, Moses-like, the soul bespeaks
The highest good its rapture seeks.

UP THE WESTERN SLOPE.

THE grandest of all the mountain ranges on the western side of the United States is the Sierra Nevada. This range from Mount Shasta, at the north, where it blends with the Coast Range, to Mount Whitney, at the south, beyond which point it breaks off into irregular formations that finally slope to the deserts, is about five hundred miles long. Its western slope, which is at least one hundred miles long on any grade fit for travel, is covered below an elevation of seven thousand feet with the most magnificent coniferous forests on the continent, embracing the wonderful groves of *Sequoia gigantea*. These forests extend to the foot-hill region, a belt of gently rounded mountains and level table-lands, where the prevailing larger growths are deciduous and evergreen oaks, the digger or nut pine, ceanothus, syringa, manzanita, buckeye, and poison-oak. The foot-hills gradually melt into the broad plains of the Sacramento and San Joaquin, only fifty or sixty feet above tide-level, which sweep their flat surfaces of emerald or golden harvests clear to the base of the purple Coast Range, rising hazy in the distant air of the Pacific. This placid region succeeds the tumultuous ruggedness of the higher ridges like a calm after a storm. Until the lower foot-hills are reached, the Sierra Nevada, on this slope, seems to break down in long, regular ridges, the outlines of which, at right angles to the trend of the range, are drawn straightly across the sky, presenting massive but precise forms, more grand than picturesque. But these ridges are divided by cañons eroded by ice and water, having a depth of one thousand to three thousand or four thousand feet, whose walls are often precipitous cliffs, and, even where clad with soil and forest, usually very steep. These cañons, with the streams which flow through them, head up in the snowy summit of the range, where they often open into meadow valleys, as the summit itself, double-crested along much of its course, holds still larger valleys, which open into the great plateau of Nevada at either extremity of the range. While the general elevation is from seven thousand to ten thousand feet, it is crowned by a multitude of peaks which reach altitudes all the way from eleven thousand to fifteen thousand feet, and on which the snow never entirely melts. Composed of splintered crags of granite, where the granite is not overlaid by even more irregularly cut volcanic rocks, the sky outline of the snowy summit is sharply serrated. Hence the Spanish name of Sierra Nevada, even more appropriate to this range than to that lesser one in Granada which originally bore it.

The comparatively timberless eastern slope of the Sierra Nevada, with its infrequent streams and monotonous gray stretches of wild sage, plunges abruptly down to the Nevada plateau. A descent of about two thousand five hundred feet, in a distance of fifty miles, is all there is of this slope, the plateau itself having an elevation of four thousand or five thousand feet and extending with its irregular mineral ridges to the Salt Lake basin and the Rocky Mountains. Thus on one side of the Sierra Nevada are verdure and fertility, — the summer charm of a semi-tropical clime, with its varied and abundant products, its poetic beauty of scenery, and its keenly sensuous joy in vitality; while on the other are barrenness and sterility, naked mountains, monotonous and often desert plains, where nature looks desperately unfinished, and gives every sign of rigorous struggle, without amenity or repose. The traveler from the east enjoys this vivid contrast so quickly realized — this rapid exchange of arid wastes for luxuriant woods and fields; but the transition going from the west chills and depresses, except at evening, when the sage-brush plains and treeless mountains of Nevada are transformed by the alchemy of color, and kindle into beauty. Probably the passage of no other mountain range of equal magnitude affords so much scenic enjoyment, at so slight an expenditure of energy, as the Pacific Railroad makes daily practicable. To know the summit of the range thoroughly, one must of course leave the railroad — must explore on horseback and afoot the wonderful gorge of Yosemite, and the equally wonderful Tuolumne cañon, with the lesser Yosemite, Hetch-Hetchy; must go to the Kern River region, where a hundred peaks rise from twelve thousand to fourteen thousand feet, and Mount Whitney soars one thousand feet higher, overtopping Shasta and every other peak in the United States outside of Alaska, unless the Colorado Mountains shall prove to contain a higher point; and must also go to Mount Lassen with its ancient crater and hot lakelet, and to the isolated cone of Shasta, most lovely and interesting of all the great peaks. But the railroad summit presents enough of the grand and picturesque, and sufficiently illustrates the character of the range, to repay a special trip, if that were the only one the tourist should make to the high Sierra.

Leaving Sacramento, rimmed about with its iron-clad levee and fringes of willow thicket, only fifty-six feet above the tide-level, the Central Pacific Railroad reaches the first swell of the Sierra within eight miles to the eastward, and in one hundred and five miles makes the summit in Donner Pass, seven thousand and forty-two feet above the sea. In the spring — say from February or March to June — a trip to the summit is especially striking for the sharp contrast between the Eden-like beauty of the lower country and the Arctic pallor of the region within the snow-belt. The plains of Sacramento, where they are not broken with the plow or sown with grain, are covered with a profuse growth of many-colored wild flowers, most brilliant of which is the California poppy (*Papavera Eschscholtzia*), whose deep orange cups flame out in sunny splendor where they are massed in large tracts, and are seen in glowing contrast by patches of blue lupin and larkspur. On this gay parterre flourish at intervals park-like groves of large oaks, deciduous and evergreen, with huge bunches of mistletoe tangled in their leafy tresses, their gray trunks circled at the base with flowers that court their shade, recalling the myth of the fairy dancing-ring. The common brown meadow-lark, and the equally plain linnet, make these gay scenes vocal with unfailing song. The atmosphere is singularly clear and pure; the sky a soft and tender blue, suggestive of infinite space; the whole influence of the landscape and the season intoxicating. And the floral profusion extends to the rolling foot-hills, albeit the reddish tint of the soil shows through its vernal dressing, and a few lowland pines begin to dispute the sway of the smaller oaks. The ceanothus, or California lilac, with its honey-breathing bloom, here comes in a thick underbrush, mixed with the manzanita, whose smooth limbs are as red as a cherry, and whose thick leaves are as stiff as wax. Groups of buckeye rise in higher masses of lighter green, relieved by spikes of small blossoms, that bristle all over them.

This is pretty much the character of the Sierra foot-hills up to the edge of the snow-line, say twenty-five hundred to three thousand feet above the sea.

where begin the coniferous woods which stretch up to the base of the highest peaks, and where one gets the first fine outlook over the wild valleys below, the first glimpses of blue cañon on either side, and the first view of those long straight ridges which lead up to the crest of the range by gradations at once easy and stupendous. During the spring months the vegetation between the plains and the coniferous belt is very bright and fresh, and there is no dust. Later, when shrub and grass are dry and russet-colored, and the red or brown soil rises in clouds, making the hot air oppressive to the traveler, there is a prevailing olive color in the underbrush, and even in the trees, especially the oaks, until the pines and firs lift their tops in lofty bowers of fresh and vivid green, carpeting the soil softly with their needles, while their cuir-colored trunks form stately colonnades, through which the sun shoots long beams of gold rayed like the chariot-wheels of Phœbus.

The portion of the lower Sierra thus far sketched is the region of the gold deposits. Here lie those great bands of slate, veined with quartz, whose degradation was the source of the precious metal distributed through the overlying drift, in the channels of modern streams, in the beds of ravines, and on the summits and slopes of hills. Here the chocolate-colored rivers, choked for a hundred feet deep with mining *débris*, attest the destructive activity of the gold-hunters. Every ravine and gulch has been sluiced into deeper ruts or filled with washings from above. Lofty ridges have been stripped of auriferous gravel for several continuous miles together, to a depth of from one hundred to two hundred feet. Cataracts of mud have replaced these foaming cascades which used to gleam like snow in the primeval woods. And the woods have, alas! in too many cases, been quite obliterated by the insatiate miner. But it is pleasant to observe how nature seeks to heal the wounds inflicted by man; how she recreates soil, renews vegetation, and draws over the ugly scars of twenty years a fresh mantle of verdure and bloom. Extensive groves of young pines and cedars are flourishing on the sites of the old forests, along the course of water ditches, and even in the chasms of decaying granite and piled up boulders and cobbles left by the miner. Small basins and valleys once covered or filled with mining litter are coating over with grass and grain, and in some instances have been converted into garden spots. Indeed, many of the old mining camps are now more noted and valuable for their orchards and vineyards than for their gold product. The rude log-cabin has given way to the vine-clad cottage, and the oleander blooms before doorways where once the only shrub may have been the pretty but noxious poison-oak. Coloma, where gold was first discovered in 1848, and where five thousand men dug for it once, is now a sleepy little village of horticulturists and vintners, embosomed in sloping hill-side vineyards, its "saloons" abandoned to the rats, and its jail converted into a wine-cellar. On the very verge of deep hydraulic diggings cling thrifty orchards. The peach, the fig, and the prickly pear are rivals in luxuriant bearing, clear up to the line of winter snow, and even the orange grows, where it has been tried, two thousand five hundred feet above the sea. Ditches cut at great expense to bring water to the diggings now serve to irrigate gardens, orchards, and vineyards. Even the rapidly passing railway traveler catches suggestive glimpses of all these changes, betokening the new era of permanent settlement and culture which is coming to the rude places of old.

Yet it is a relief to get out of sight of the crater-like chasms left by the miner, with their pinky chalk-cliffs of ancient drift, along which the cars fly as over a parapet or wall. It is pleasant to quit the hills denuded of timber and left so desolate in their dusty brown; delightful to reach loftier ridges and plunge into cool shades of spicy pine. Here nature seems to reassert herself as in the time of her unbroken solitude, when the trees grew, and flowers bloomed, and birds caroled; when the bright cataracts leaped in song, and the hazy cañon walls rose in softened grandeur, indifferent to the absence of civilized man; though the civilization which has made these superb heights so easily accessible for our enjoyment is not to be scorned. The rocky promontories, jutting into blue abysses, and giving sublime pictures of mountain lines sweeping down to the plain, are finer for the iron rail which lies along their dizzy edges, surpassing the Appian Way of the Romans, or the Alpine Road of Napoleon. Here we have the sensation of ballooning without its dangers. Flying over deep gulches on trestles one hundred feet high, and along the verge of cañons two thousand feet deep, we look out on the air and view the landscape as from a perch in the sky. Thus is the picturesque made easy, and thus mechanical genius lends itself to the fine wants of the soul. Reaching the deep snow-belt, however, the vision of mountain scenery is cut off by the many miles of snow-sheds, or, at best, is only caught in snatches provokingly brief, as the train dashes by an occasional opening. If the time is winter, the shed is enveloped in snow from ten to twenty feet deep; the light gleams feebly as through diaphanous shell, and the smoke-blackened interior is in sharp contrast to the white drifts seen through chinks and slits. A ride through these winding galleries at this season is weird enough, and the rare glimpses without reveal a scene thoroughly arctic. The woods are grand with their drooping plumes, — white on the upper, green on the lower surface, — and the massive trunks are clad on one side with a thick garment of greenish-yellow moss extending to the limbs, which often trail long pendants of gray or black moss from bark or foliage. Higher up, the treeless peaks and slopes of granite, dazzlingly white, send down roaring torrents. The sea-murmur of the forest has ceased; there is a hush in the air except for the roar of waters. The cushion of snow prevents reverberation, and muffles the harp of the summer-sounding pine. Here and there in the sheds are cavernous side-openings, which indicate snow-buried stations or towns, where stand waiting groups of men, who receive daily supplies — even to the daily newspaper — in this strange region. The railroad is the raven that feeds them. Without it these winter wildernesses would be uninhabitable.

When the train has passed, they walk through snow-tunnels or smaller sheds to their cabins, which give no hint of their presence but for the shaft of begrimed snow where the chimney-smoke curls up. And in these subnivean abodes dwell the station and section people, and the lumbermen, during several months, until the snow melts and its glaring monotony of white is suddenly succeeded by grass and flowers, except where the granite crests hold the snow longer, and seldom bear richer vegetation than lichens and a few straggling dwarfs of pine or cedar.

Nothing can be more charming than the woods of the Sierra summit in June, July, and August, especially in the level glades margining the open summit valleys, at an elevation of from six thousand to seven thousand feet. The pines and firs, prevailing over spruces and cedars, attain a height ranging from one hundred to two hundred feet, and even more. Their trunks are perfectly straight, limbless for fifty to a hundred feet, painted above the snow-mark with yellow mosses, and ranged in open park-like groups, affording far vistas. The soil may be thin, but it is soft and springy to the tread, covered with needles of the pine, greened with tender grasses and vines, and thickly sprinkled with blossoms. Huge boulders of granite relieve the vernal coloring with their picturesque mosses of gray, starred with lichens. These rocks are often hid in vines or in dwarf oaks and manzanitas, which, under the pressure of deep snow, assume a vine-like growth, winding about a boulder with their clinging and sinuous small branches. Thickets of wild rose and other flowering shrubs occur at intervals, giving an almost artistic variety to the woodland scene. The crimson snow-plant lifts its slender shaft of curious beauty. Large patches of helianthus — some species with very broad leaves — spread their sunflowers to the air. Sparkling springs, fresh from snowy fountains, silver-streak these forest meadows, where birds come to bathe and drink, and tracks of the returning deer are printed. Once more the quail is heard piping to its mates, the heavy whirring flight of the grouse startles the meditative rambler, and the pines give forth again their surf-like roar to the passing breeze, waving their plumed tops in slow and graceful curves across a sky wonderfully clear and blue. To the citizen weary of sordid toil and depressed by long exile from nature, there is an influence in these elevated groves which both soothes and excites. Here beauty and happiness seem to be the rule, and care is banished. The feast of color, the keen, pure atmosphere, the deep, bright heavens, the grand peaks bounding the view, are intoxicating. There is a sense of freedom, and the step becomes elastic and quick under the new feeling of self-ownership. Love for all created things fills the soul as never before. One listens to the birds as to friends, and would fain cultivate with them a close intimacy. The water-fall has a voice full of meaning. The wild rose tempts the mouth to kisses, and the trees and rocks solicit an embrace. We are in harmony with the dear mother on whom we had turned our backs so long, yet who receives us with a welcome unalloyed by reproaches. The spirit worships in an ecstacy of reverence. This is the Madonna of a religion without dogma, whose creed is written only in the hieroglyphics of beauty, voiced only in the triple language of color, form, and sound.

Let the pilgrim to these Sierra shrines avoid the hucksters who carry traffic into the temple. Let him leave the beaten line of travel, where the ravaging axe converts the umbrageous solitude into noisy blanks. Let him quit the scene where sawdust chokes and stains the icy streams in their beds of boulders. All things have their place, and these are well in their way, but are only an offense to the true lover of nature. Plunge into the unbroken forests — into the deep cañons; climb the high peaks; be alone a while and free. Look into nature, as well as at nature, so that the enjoyment shall be not merely sensuous but intellectual. A less exclusive and jealous pilgrimage than this, however, will make a man better, physically and mentally. He will realize from it the truth of Tyndall's testimony to the value of high mountain exercise in restoring wasted nervous energy and reviving the zest and capacity for brain work. He will find in it a moral tonic as well, and come back to the world, not loving men better, perhaps, but more patient and tolerant, more willing to accept work with them as being itself better than the thing worked for.

SUNRISE NEAR HENNESS PASS.

The moon is streaming down her mellow light
Upon the snowy summits of the range
That walls apart the gold and silver lands.
It gleams in piney glens and cañons wild,
On tumbling cataracts and singing rills,
That are not seen but heard amid the gloom;
Making the savage scene, remote and lone,
Seem holy as the fane where thousands kneel
And worship 'neath the dome that Art hath reared.

Is that a rival moon whose tender glow
Now silvers in the east the speary points
Of bulky pines that crowd the mountain pass?
No, it is Venus, prophet-star of day —
The lovers' planet, lambent, large, and full;
And what a lunar glory trails she now
Along the dewy chambers of the morn!

But moon and planet pale and dwindle small
Before the coming of a greater orb;
As eyes of love, that brightly beamed in life,
Contract and darken 'neath the glare that streams
Upon them from the realms of fadeless light.
The gray sky whitens with a boreal glow
Along the farthest dark blue line of hills;
Then flushes into amber faint, and then
To saffron hues that kindle into rose.
Life stirs with dawning light. The birds awake,
And welcome it with twitterings of joy,
Hoarse murmurs from the Yuba's fretted stream
Come faintly up from depths of gorges dark.

The cool air, rising over banks of snow,
With gentle rustling fans the cooing birds;
And all the dusky woods are stirred and thrilled
With swelling of the Memnon strain that flows
From touchings of no priest but Nature's self.
Peak after peak beacons the coming day,
And snowy summits blush like maiden cheeks
At nearing footsteps of expected swain.
The splintered pinnacles and rocky crags
That late frowned gloomily as castles old

Perched on the dizzy heights that guard the Rhine,
Now softly rise in gold and purple air,
And move the soul like sad and stately verse.
The east is all aglow with brightening flame,
That overflows the willow-fringèd vale,
And drives the shadows from ravine and glen.
In ghastly pallor wanes the rayless moon,
While jeweled Venus has evanished quite.

Oh, what a burst of splendor! A great globe
Of burning gold, flashing insufferably,
And warming all the scene with ardent ray,
Heaves into view above the mountain's line,
Darts golden arrows through the dusky aisles
Of thickly-columned cedar, pine, and fir,
Transmutes the common dust to shining haze,
Licks up the rising mists with tongue of flame,
Gilds the "pale streams with heavenly alchemy,"
And down the shaggy slope, for scores of miles,
Pours forth a cataract of tremulous light
That floods the valley at its rolling base,
Making the arid plain a zone of tropic heat.

ON THE SUMMIT.

Arrived at the summit of the Sierra Nevada, on the line of the railroad, there are many delightful pedestrian and horseback excursions to be made in various directions. At Summit Valley (which is associated with the relief of the tragically fated Donner emigrants, and is only three miles from Donner Pass) there is an odious saw-mill, which has thinned out the forests; an ugly group of whitewashed houses; a ruined creek, whose waters are like a tan-vat; a big sandy dam across the valley, reared in a vain attempt to make an ice pond; a multitude of dead, blanched trees; a great, staring, repellant blank; and yet this valley is not unlovely. Its upper end, still a green meadow, leads to the base of peaks ten thousand or twelve thousand feet high, whose light gray summits of granite, or volcanic breccia, weathered into castellated forms, rise in sharp contrast to the green woods margining the level mead. A little apart from the noisy station the woods are beautiful, as we have described them, and the boulder-strewn earth reminds one of a pasture dotted with sheep. On the northern side rises the square butte of Mount Stanford, two thousand four hundred and fifty-three feet above the valley, and nine thousand two hundred and thirty-seven feet above the sea. Its volcanic crest is carved into a curious resemblance to a ruined castle, and hence it was named, and is still popularly called, Castle Peak; but as the same title is affixed to several peaks along the range, the state geologist has wisely given it another on the official maps. This peak can be ascended to the base of the summit crags on horseback; the remaining climb afoot, up a very steep slope of sliding *débris*, is arduous but short, and is repaid by a superb view, embracing at least a hundred miles of the Sierra crests, their numerous sharp peaks streaked with snow, and lying between them at intervals the many lakes of the region, including the flashing sheet of Tahoe, nearly thirty miles long, the dark and deep-set Donner, and the little meadow-fringed lakes of Anderson Valley; while on either side stretch the slopes of the range, rugged, with vast exposures of granite, overlaid here and there by the lava of ancient craters, and bristling lower down with receding coniferous woods, that melt into the purple distance as the ridgy flanks of the range sink at last into the hazy plains. On one side of this characteristic peak the foot-climber stops to rest on a depression where grass and flowers grow luxuriantly, and swarms of humming-birds hover over the floral feast, their brilliant iridescent plumage flashing in the sun, and the movement of their wings filling the air with a bee-like drone. Above all this beauty frown the bare volcanic cliffs and pinnacles that top the mountain-Eden and the desert side by side. The upper Sierra is full of contrasts and surprises. After tedious walking over rocky barrens, or toilsome climbing up slippery gorges, in the very path of recent torrents, one comes suddenly on little bits of garden and wild lawn, where butterfly and bird resort, and the air is sweet with perfume. At the base of cliffs which looked forbidden at a distance, cool springs will be found, painting the ravines with freshest green; red lilies swing their bells, lupins and larkspurs call down the tint of heaven; ferns shake their delicate plumes, bright with drops of dew; and the rocks offer soft cushions of moss, the precipice above, where water trickles down, being clad with lichens, and a hundred crannied growths. The delighted pedestrian lingers at such oases, loath to go forward. Goethe says, "Great heights charm us; the steps that lead to them do not." But this is hardly true in a great part of the Sierra Nevada, where the scenery by the way lightens the labor of climbing, and the sensation at the summit is only the climax of protracted enjoyment.

The tourist who stops a few days at Summit Valley will find a walk along the railroad, through the snow-sheds, peculiarly entertaining. These sheds, covering the track for thirty-five miles, are massive

Section of Snow-shed.

arched galleries of large timbers, shady and cool, blackened with the smoke of engines, sinuous, and full of strange sounds. Through the vents in the roof and the interstices between the roof-boards, the sunlight falls in countless narrow bars, pallid as moonlight. Standing in a curve the effect is precisely that of the interior of some old Gothic cloister or abbey hall, with the light breaking through narrow side windows. The footstep awakes echoes, and the tones of the voice are full and resounding. A coming train announces itself miles away by the tinkling crepitation communicated along the rails, which gradually swells into a metallic ring, followed by a thunderous roar that shakes the ground; then the shriek of the engine-valve, and in a flash the engine itself bursts into view, the bars of sunlight playing across its dark front with kaleidoscopic effect. There is ample space on either side of the track for pedestrians to stand as the train rushes past, but it looks as if it must crush everything before it, and burst through the very shed. The approach of a train at night is heralded by a sound like the distant roaring of surf, half an hour before the train itself arrives; and when the locomotive dashes into view, the dazzling glare of its head-light in the black cavern, shooting like a meteor from the Plutonic abyss, is wild and awful. The warning whistle, prolonged in strange diminuendo notes, that sound like groans and sighs from Inferno, is echoed far and long among the crags and forests.

Summit Valley, lying three miles west of the highest point on the railroad, is six thousand seven hundred and seventy-four feet above the sea. The air is keen and invigorating; there are few summer nights without frost, but the days are warm enough for health and comfort. Nine miles southward, and six hundred and sixty-one feet lower, are the little known but remarkable "Summit Soda Springs." The drive to these springs is one of the most picturesque and enjoyable in the Sierra. Passing by fine dark cliffs of volcanic breccia to the right, and over low hills covered with tall, red firs, the road leads to Anderson Valley, a green meadow, embosoming three little lakes, which are perfectly idyllic in their quiet beauty. These lakes are the remnants of a larger single body which evidently once filled the whole valley. Their outlet is through a narrow rocky gorge which empties into a tributary of the north fork of the American River. The road follows the steep side of this gorge for a short distance, then reaches the summit of a ridge overlooking the cañon of the American, two thousand feet below. Looking down this cañon, one sees rising from its blue depths the grand bulk of Eagle Cliff, — a rocky promontory whose top is probably eight thousand feet above the sea, and whose bald slope to the river presents a precipitous front of inaccessible steepness. The largely exposed mass of this elevation makes a magnificently long outline across the sky, and when the cañon is hazy in the afternoon, and the sun declines towards the west, the sharp sculpture of the cliff is obscured behind a purple veil and presents a front of ethereal softness, like a vast shadow projected against the heavens, or a curtain let down from the infinite. Directly across the cañon, looking southward, the ridge separating the north American from the middle fork of the main river sweeps up in a still longer and grander line, which swells into snow-peaks from nine thousand to ten thousand feet high, — as high above the valley at the bottom of the cañon as Mount Washington is above the sea, — exposing four thousand feet of uplift to the glance, and weathered into a rich variety of pinnacled, domed, and serrated forms. The descent into the cañon is a long zigzag through a lovely forest, in which the red fir, with its deeply corrugated bark, attains a height of from one hundred and fifty to two hundred and fifty feet, and frequently has a thickness at its base of four or five feet. The yellow pine (*P. ponderoso*), even more massive, lifts its rich foliage above a bright and leather-colored trunk, the bark on which is almost smooth, and is divided into long plates. But the monarch of these woods (though infrequent here) is the sugar pine (*P. Lambertiana*), whose smooth trunk, often six feet through, rises a hundred feet or more without a limb, perfectly straight, and is crowned with a most characteristic, irregular, and picturesque top, its slender cones, a foot or more in length, hanging from the tips of the boughs like ear-drops. The eye constantly seeks out these magnificent trees, and every large one is hailed with admiring exclamations. Dwarf oak and manzanita, ceanothus and chemisal, are the prevailing underbrush. In sunny open spaces, or on bits of timberless meadow, the rose, and thimble-berry, and a purple-blooming asclepia abound. Occasional large patches of a broad-leafed helianthus, when not in bloom, curiously resemble ill-kept tobacco fields. About grassy springs a very fragrant white lily sparingly unveils its virgin beauty. A spotted red species of the lily is more common, and small, low-flowering plants are numerous. The southern slope of the ridge, descending to the soda springs, has a deep soil and is very thickly timbered. At its base the small streams are lined with thickets of quaking aspen, cottonwood, and balm of Gilead, alternating with more continuous groves of alder and willow, where the prevailing undergrowth is a silkweed, four or five feet high, whose slender stalks, bearing narrow, sharply-cut leaves, are thickly crowned with purple blossoms. Thickets of thorn afford cover for numerous quail. Coniferous trees continue along the narrow banks of the river, but stand more apart. At the head of the cañon, the granite breaks down in huge benches, or shelves, presenting perpendicular faces as looked at from below. The river tumbles a hundred feet, in cascades and falls, through a gorge of granite set in a lovely grove of cedar and pine, — and pools of green water sparkle in clean basins of granite at the foot of every fall. The rock of this gorge is richly browned and polished, except on the gray faces of the cliffs overhanging the stream. Farther up the cañon, where the main crest of the Sierra describes the arc of a circle along the eastern sky, and is crowned by several high peaks, the granite is overlaid with lava and breccia, the product of the volcanoes which anciently dominated and overflowed this region, and whose relics are seen in the sharp cones of trachyte at the summit. Near the junction of granite and volcanic rocks, numerous soda springs boil up through seams in the ledges, often in the very bed of the stream. The water of these springs is highly charged with carbonic acid, is delightfully cool and pungent, and contains enough iron to make it a good tonic, while it has other saline constituents of much sanitary value. Where the fountains bubble up they have formed mounds of ferruginous earth and soda crust, and their water stains the river banks and currents at intervals. One of the largest and finest springs has been utilized, forming one of the most picturesque resorts in California. About two miles below, the river has cut a narrow channel one hundred and fifty feet deep and one eighth of a mile long through solid granite. This chasm is but a few rods wide at top, and only a few feet wide at bottom, where there are numerous smooth pot-holes, forming deep pools of wonderfully transparent water of an exquisite aquamarine tint. There is enough descent to make the current empty from one pool to another in little cascades, over sharp pitcher-lips of polished rock. Lovers of angling are provoked to find no fish in these charming basins. A few stunted but picturesque cedars are stuck like cockades in the clefts above, and the summits of the chasm walls are rounded and smoothed by ancient glacial action. To this place was given the name of Munger's Gorge, by a gay picnic party last summer, in honor of the fine artist who sat with them on its brink, and was first to paint it. A few miles below is a still deeper and grander gorge, at the foot of Eagle Cliff, where the precipitous granite walls rise a thousand feet or more, and the stream makes a sheer fall of a hundred feet. Above this fall fish cannot ascend, and so it happens the beautiful upper river is the angler's disappointment. There are many fine climbs to be made in the vicinity of the soda springs, including Mount Anderson and Tinker's Knob, companion peaks, separated only by a saddle-like depression a few hundred feet deep and scarcely a mile long, at the very head of the cañon, dividing it from the head of Tuckee River, on the eastern slope, by a few miles. These peaks, having an elevation from three thousand to three thousand five hundred feet above the river, and from nine thousand to nine thousand five hundred above the sea, can be climbed with comparative ease in a few hours. Tinker's Knob, the higher of the two (named after an old mountaineer, with humorous reference to his eccentric nasal feature), is a sharp cone of trachyte, rising above a curving ridge composed partly of the same material and partly of lava and breccia overlying granite. Its summit, only a few yards in extent, is flat, and paved with thin slabs of trachyte, and cannot be scaled without the aid of the hands in clambering over its steep slopes of broken rock. Anderson is shaped like a mound cut in half and is composed of breccia (volcanic conglomerate), rising on the exposed face in perpendicular cliffs, similar to those which occur lower down the slopes. The ridge crowned by these twin peaks is approached over a steep mountain of granite boulders, morainal in character, which leads to a tableland clad sparsely with yellow pines and firs. Clambering over the broken rock to the top of Tinker's

Knob a magnificent panorama is unfolded. Over three thousand feet below winds the American River, — a ribbon of silver in a concavity of sombre green, seen at intervals only in starry flashes, like diamonds set in emerald. The eye follows the course of the cañon fifty or sixty miles down the western slope, marking the interlapping and receding ridges which melt at last into the hazy distance of the Sacramento Valley. With the afternoon sun lighting up this slope, shooting its rays through the ranks of pines, and making glorious the smoke of burning forests or the river vapors, which soften without concealing the scene, the effect is wonderfully rich. Looking north and south, the eye discerns a long procession of peaks, including Mount Stanford, the Downieville Buttes, and Mount Lassen. To the east lies Lake Tahoe, revealed for nearly its whole length, with environments of picturesque peaks. There, too, lies its grand outlet, the basin of the Truckee River, which can be followed for fifty miles to the Truckee meadows in Nevada, past several railroad towns. The line of snow-sheds from the ridge above Donner Lake to Truckee is distinctly seen, and the roar of passing trains comes faintly up. The Washoe Mountains bound the view in that direction, completing a grand picture. The view is amphitheatrical, and the radius of it cannot be under two hundred miles.

A still finer outlook can be obtained from a somewhat higher peak to the southward, which heads the next cañon in that direction, and is approached over or along a succession of volcanic spurs, edged with sharp cliffs of breccia, of true drift conglomerate, and narrow plateaus of the same material resting on vertical walls of basalt. The cliffs in one place are a dark Vandyke brown, faced with brilliant red and yellow lichens, and the talus at their base is a grassy slope of vivid green. Opposite these, across a gulf perhaps two thousand feet deep, rises the bluff face of the peak we seek,—shaped like the South Dome of Yosemite, but a mass of crumbling breccia of a pale chocolate or drab color, enameled with patches of snow. Some hard climbing is necessary to surmount this, but the view repays the labor. Though much of the character described above, it is more extensive, giving a finer idea of the summit peaks for a distance of one hundred and fifty miles along the range. Mount Lassen and the Black Butte, its neighbors, — volcanic cones both, — are beautifully exposed, and towers higher than any mountain points in that direction until Mount Shasta is reached, only seventy miles farther north. Looming into view one after the other, as the eager climber ascends, they excite the mind and stimulate the weary limbs to renewed effort; and as the view, at first limited by near ridges, expands to a vast circle, melting on every side in the atmosphere, the soul expands with it, and the very flesh that holds it grows buoyant.

"What now to me the jars of life,
 Its petty cares, its harder throes?
The hills are free from toil and strife,
 And clasp me in their deep repose.

> "They soothe the pain within my breast
> No power but theirs could ever reach;
> They emblem that eternal rest
> We cannot compass in our speech."[1]

A couple of thousand feet below are several little blue lakelets, fed by melting snows, in small basins of verdure. Flowers bloom in gold and blue and purple beauty at their margins, and at the very edge of the frozen snow. A fitful breeze sweeps a quick ripple of silvery wrinkles over the placid pools, and they are smooth and blue again in an instant. There is no cloud in the sky, but shadows of high-flying birds pass over the landscape below, reminding us of clouds, and intensifying the sensation of vast space and depth. Recovered from the ecstasy of this grand scene, we begin to study the geology of the region, which is beautifully revealed. First, an upheaval of granite, rupturing, displacing, and metamorphosing the beds of sedimentary rock deposited when the ocean lay over the sight of the range.

[1] John R. Ridge.

Then, over the granite, and crowning all the highest ridges and peaks, are layers of volcanic rock — trachyte, breccia, red lava, pumice, and scoria — cut through clear to the underlying granite at the head of cañons, first by the glaciers that succeeded the volcanic period, and later by frost and freshet, by slides and avalanches. The evidences of glacial action below the long line of ancient craters, can be clearly traced in the excavation of the lava flows; in the rounded and polished masses of granite; in the erratic boulders left here and there, perched like monuments on solid ledges; in the morainal deposits cut through by modern streams or still forming lakes. Thus the reign of ice succeeded the reign of fire, and both these tremendous forces were needed to fashion the rich mountain forms, and to prepare the way for all the lovely forests greening their flanks.

Perhaps a little finer exhibition of glacial action is that to be seen in the cañon of the South Yuba, leading out of Bear Valley, twenty-two miles west of the railroad summit, and a little north from the Soda Spring region. Bear Valley is about a thousand feet below the ridge along which the railroad passes. It was anciently filled by a lake caused by the terminal moraine of a glacier. The cutting through finally drained the lake, and left, first a morass, then a meadow. Going up the valley two or three miles, to the mouth of a deep gorge, the observing traveler will notice many glacier-polished hills of granite — bare mounds of rock that were carved into shape by a moving body of ice, ages ago. The gorge itself has been cut down to a depth of from three hundred to eight hundred feet through granite; and its walls, curved and sloped at their summits, and sharply cut and polished on their faces, frown over the stream that drops from one green bowl of rock to another at their clean-swept bases. Immense pot-holes, still retaining the boulders that excavated them, are frequent through the bottom of this wild gorge. Some of them have been worn through on one side and form little cascades.

For the purpose of conveying the pure water of the Yuba to Nevada City a narrow flume covered with planks has been built through this gorge, which would else be inaccessible to the tourist. Over this pathway one can walk into the rocky chasm for two miles. The construction of the flume was a work of difficulty and danger. It is supported partly by walls laid up on the outer side; partly by iron bars and wire cables fastened in the solid rock, which hold it in suspension over perpendicular depths. The face of the rock had to be blasted to make way for it, and the blasting could be effected in places only by letting men down from the top of the cliff with ropes, and they drilled and charged the powder-holes, hung in mid-air. One poor fellow, who put off a blast prematurely, was blown from his airy perch across the river and dashed in pieces. Walking securely along this flume, one looks down a sheer precipice into the yawning river-holes far below, enjoying their transparent green and the snowy play of their cascades, and wondering at the force which cut those enormous bowls in the solid granite, and which keeps the whole bottom of the gorge swept clean and smooth. Looking up, on one hand, the neck stiffens and the eye wearies with the effort to see the whole of the perpendicular cliff. The lofty coniferous trees above, which sometimes nod over the beetling edge, are dwarfed by the distance. The face of the cliff is moist here and there with dripping springs, which cover it with exquisite mosses and many rare flowering plants, ferns, and vines, the delight of botanists. The less erect wall on the opposite side, scarcely a stone's-throw across, is brown and gray with motley lichen patches. It is a place to linger in for hours, and to leave with regret.

Returning to the summit, let us leave the railroad at the point where it begins its descent of the eastern slope, and climb the tree-covered ridge and bald granite cliffs overlooking it to the left. A thousand feet above the pass will give an elevation about eight thousand feet above the sea, commanding a view of

CROWN OF THE SIERRA.

Donner Lake and the valley of the Truckee, over two thousand feet below, and down the eastern slope to the transverse mountain lines of Nevada, sixty miles off. Right and left the view is obstructed by crags and pinnacles of bare granite, which loom up cold and gray against the intense blue, except when the morning or evening light warms and empurples them, or tinges them with rose, as seen afar in the last glow of sunset. Among these rocky summits lies Lake Angela, gemmed in the granite and girdled with fir groves and narrow fringes of grass and flowers, — a cup of stone, decorated on its sides with Nature's own graceful arabesque. Donner Lake is sunk in a narrow, oblong cañon, cut through the granite by one of the ancient glaciers of the eastern slope, a tributary, probably, of the enormous ice-river which once put out of the basin of Lake Tahoe and occupied the present channel of the Truckee. The descent to Donner from the granite peaks at its western end is abrupt and rugged, and the view from those peaks is remarkable for its stern grandeur. It was near this point that Bierstadt made the studies for his most faithful picture of California scenery. At

Donner Lake, Mount Sanford in the distance.

the base of the cliffs the lake, an irregular oval three miles long, and half a mile to a mile wide, steel-gray or dark lead in color, when the sun is not flashing from its smooth surface, or the silvery vapors are not rising, framed by sloping ranks of spear-headed pines; beyond the lake, a dark trough ending in a sky-line of lofty mountains, softened by the pearly gray of morning, and exposed in all the sharpness of their rocky anatomy by the glow of evening, which tints them a color the despair of art, — this completes the picture of Donner.

But the gem of all scenes in this part of the Sierra is Lake Tahoe, situated about fifteen miles southerly from Donner, between the double crests of the range, measuring about twenty-three miles long from northeast to southwest, by about fifteen miles wide at its widest, having an altitude of six thousand two hundred and eighteen feet above the sea, and being surrounded by mountains that rise from one thousand to four thousand feet higher, volcanic for the most part, except in the southwest, where they are granitic. The favorite road follows for fifteen miles the banks of its outlet, Truckee

River, — a rapid stream of remarkably clear water, having a width of from sixty to a hundred feet, and flowing over a bed of boulders, between groves of alder, willow, maple, cottonwood, and aspen. The heavily timbered ridges, putting down in nearly straight lines from the summit, rise on either side of this stream to a height of from one thousand to two thousand feet, at a sharp angle, and are composed of volcanic rock, originating with the extinct craters of the crest, and sometimes exposed in high and picturesque cliffs of a rich color. Extensive logging operations are conducted along the Truckee, and it is one of the sights of the trip to witness the shooting of the logs along timber-ways for one thousand two hundred feet down the side of the ridge. They make the descent in thunder and smoke, and each log, as it strikes the water, will send up a beautiful column of spray a hundred and fifty feet, resembling the effect of a submarine explosion. The banks of the river are strewn with granite boulders and cobbles, which could only have been brought from the head of the lake by a glacier, since the adjoining ridges are entirely volcanic clear down to the stream. Indeed, glacial marks are plain enough on the rocks about the lake, the polish even remaining on one exposure of volcanic rock on the eastern shore near Tahoe City. Imposing as must have been the Tahoe or Truckee glacier, it was narrower below the present lake-bed than one of three glaciers still living on the flanks of Mount Shasta, — the Agassiz Glacier, as named by Clarence King, — which has a width of about three miles; whereas the Truckee is hardly so wide as the Whitney Glacier, — about half a mile. The first sight of the lake is very striking as one breaks from the sombre-hued forests of pine and fir, and gazes on a wide expanse of blue and gray water, sparkling in the sun, and relieved by a distant background of violet-colored mountains. There is an exciting freshness in the air, and the spirits are elate with freedom and joy. It is a treat to watch the alternations of color on the water. Prof. John Leconte, who recently made some interesting observations on this and other phenomena of the lake, says that, wherever the depth exceeds two hundred feet, the water assumes a beautiful shade of "Marie Louise blue." Where it is shallow, and the bottom is white, it assumes an exquisite emerald green color, as in the famous Emerald Cove. Near the southern and eastern shores the white sandy bottom brings out the green color very strikingly. The same authority informs us that his soundings indicate that there is a deep subaqueous channel traversing the whole lake in its greatest dimensions, or north and south. At several points in this channel the depth exceeds one thousand five hundred feet. The temperature of the water decreases with increasing depth to about seven hundred or eight hundred feet, and below this depth it remains sensibly the same down to one thousand five hundred feet. The constant prevalent temperature below seven hundred or eight hundred feet is about 39° Fahrenheit, — the point at which fresh water always attains its maximum density. The temperature of the water above the depth named was found, during the summer, to be from 41° to 67°. Owing to the above facts of depth and temperature, the lake never freezes, except in shallow and detached portions. As Professor Leconte says, before the conditions preceding freezing can occur, the water, for a depth of eight hundred feet, must cool down to 39°, for, until it does, the colder substratum will not float to the surface. The winter is over before this equalization can be effected, and so the water does not freeze. Owing also to the lower water being at a constant temperature only 7° above the freezing point, drowned bodies reaching it are not inflated by the gases resulting from decomposition at a higher temperature, and, therefore, do not float. The transparency of the water is so great that small white objects sunk in it can be seen to a depth of more than one hundred feet. Sailing or rowing over the translucent depths, not too far from shore, one sees the beautiful trout far below, and sometimes their shadows on the light bottom. It is like hovering above a denser atmosphere. But the surface of the lake easily ruffles into dangerous waves under a sudden wind, and a number of incautious persons have been lost in these cold depths which never give up their dead. The beaches of white sand, or clean, bright pebbles, rich in polished agate, jasper, and carnelian, margined with grassy meads where the strawberry ripens its luscious fruit, and running close to park-like groves of pine, fir, and cedar, afford delightful rambles. The shore-lines are informal and picturesque, opening into green coves and bays, where sometimes a cascade comes foaming down from the snow-peaks, or pushing out sharp points of timber and long strips of reedy marsh, leading to valleys where smaller lakes are found glassed amid a close frame-work of rocky heights. One of the prettiest of these side lake-lets rejoices in the poetic name of Fallen Leaf Lake, from the circumstance that its placid surface is often strewn with the leaves of deciduous trees blown from the banks. Another is called Cascade

Lake, because a little water-fall tumbles over a ledge into its bosom. Both of these small sheets have often been painted by the artists who repair to Tahoe every summer; but their favorite is the large lake, with its superb mountain boundaries, which on the northwest are lofty, snow-clad, and beautifully sculptured. The afternoon haze over mountain and lake is a delicate, pearly gray. Later, this color shades off into violet, and, as the sun sinks, the mountains take on the most delicious crimson flush, deepening into purple, while the lake is wonderful in its play of reflected color, and at a certain hour looks like an opal set in rubies. The moon at night converts the surface into a shield of flashing silver. By day or night the musical lapse of the wavelets on the beach charms and soothes; and when all the solitude of its original loneliness seems to come over the scene again, we have the sensation of an awful spiritual presence, "felt in the heart and felt along the blood."

EL RIO DE LAS PLUMAS.

River of feathers — calm and graceful stream!
    They name thee well. The yellow willows droop
Toward thy tranquil face like plumes, and seem
    At times to kiss thee, as fond lovers stoop
To kiss the eyes that mirror back their own.
    And in a line of beauty gently flows
Thy winding water, to the world unknown, —
    The sordid, plodding world, — but not to those
For whom the river or the brook hath all
    The wonder of old ocean's stormy flood,
Whose minds see beauty in the leaves of fall
    As in eve's fleecy cloudlets dyed in blood.

To such, dear stream, thou hast a charm; the flash
    Of silver lightning from thy glassy face.

Inclosed by foliage like a lake, and dash
    Of thy broad, foaming rapids, have a place
Alike in admiration's seat, and fix
    Upon the often grieved and grieving mind
Those recollections of delight that mix
    And brighten others of a darker kind.
For all the beauty of a lovely scene
    Beams not upon the eye to live no more
Than while we gaze : ah no!   Its spell serene
    Sinks in the heart for aye, and when we pore
In after years o'er mem'ry's tinted page,
    That lovely landscape rises to the view,
Attired in all the charms of early age,
    And seems our primal joyance to renew.

Again we see the triple peaks that rise
    Like purple isles above the yellow grain,
As lonely 'gainst the deep and cloudless skies
    As are the pyramids on Egypt's plain.
Here, in a park-like grove of mighty oaks,
    Whose trunks are crimson with the poison vine,
The acorn-hiding bird, with rapid strokes,
    Startles the echoes where the deer recline.

Afar we hear the laugh of Indian girls,
    Or murmur of the red man's alder flute :
Their camp smoke floats away in pallid curls :
    The breeze sinks low, and then the air is mute,

Save that at eve we hear the cricket's keen
    And quiv'ring music, or the hollow note
Of water-bubbling frog, and catch between
    The turtle's plaint, low in his feathered throat ;
Or listen to the hooting of the owl, —
    The ghostly owl, that only stirs at night,
When darkness wraps the landscape like a cowl
    And superstition shudders with affright.
But here are peace and love, that brood alway
    In blessed calm above a witching scene ;
And here the soul, a flower that shuns the day,
    Opes to the night and feels a joy serene.

HEAD-WATERS OF THE SACRAMENTO.

The upper Sacramento Valley is a vestibule that leads to the high altar of Mount Shasta. At first, a broad, level plain, — so broad that the Coast Range and Sierra Nevada, on either side, are but dimly seen, low in the hazy horizon, — it narrows going northward, until its mountain walls, drawing nearer and nearer together, intermix at last, leaving only a channel for the waters of the Sacramento River, lying between high and steep ridges parallel with its course for seventy miles, and then opening into a series of small valleys, at a considerable elevation, encircled by loftier mountains, where burst forth the springs that feed the river and its branches. Dividing several of these small valleys, at the very head of the

Sacramento, rises the noble bulk of Shasta, a landmark to the traveler in the great valley below for a hundred miles or more, and visible from high points to the southward for quite two hundred miles, — a snowy cone projected against the sky, without a rival peak. To the pedestrian or horseman, who makes his way slowly toward this landmark, it is a guide and an inspiration for days. In the early times, when the great valley was one wide field of flowers in the spring, or a rippling sea of wild oats in the summer, the distant aspect of the mountain, through the wonderfully clear atmosphere of this climate, and in contrast with so much vernal color, was peculiarly fine. Many a pioneer gold-hunter retains still, in whatever different and remote scene he may now be, the vivid impression of its beauty. And even yet the approach to Shasta is full of allurement, at the beginning of summer, when green and flowery tints prevail, and before the smoke of forest fires has spread an obscuring haze through the sky. At this season the valley itself is enjoyable for its verdure and brilliant

MOUNT SHASTA, FROM CASTLE LAKE

bloom; for its clean, open groves of large oaks; for its denser timber-lines along the dry channels of winter streams; for its gradual upheaval into the mound-like swells that prelude the foot-hills; for the cool, sharp vision of Sierra snow-crests to the eastward, and the lower and softer wall of purple which marks the Coast Range. The Sacramento River winds slowly its dark greenish current, at first between low banks fringed with brier and grape thickets, overtopped with sycamores, alders, willows, and cotton-woods; then between bluffs of clay or gravel, where the undergrowth is missing. Over the wide, level surface, in some directions, there is not a tree to break the monotony; but along the horizon, on warm days, are cheating visions of trees and water. It is a relief to strike the oaken parks again, and to see the mountain chains drawing closer. Here at the right stands Mount Lassen, dominating this portion of the Sierra, though only the centre of a colony of ancient volcanoes, whose crater-cones have an elevation ranging from nine thousand to nearly eleven thousand feet. From the summit of the highest peak on Lassen, in the clear season, a view is obtained extending from Mount Hamilton, in the Coast Range below San Francisco, to Mount Pit, in the Siskiyou region at the north, a distance in a direct line of nearly three hundred and fifty miles; while the view east and west extends from Pyramid Lake, in Nevada, to the coast ranges overlooking the Pacific.

At the point where Mount Lassen is most plainly seen from the valley, the foot-hills of the interblending ranges are distant only a few miles, and to this point the traveler can now go from Sacramento by rail, in the cars of the Oregon division of the Central Pacific Railroad — distance one hundred and seventy miles. The next seventy-five miles of the journey, to the foot of Mount Shasta, is made in one of the stages which runs through from Redding, the railway terminus, to Roseville, the southern terminus of the railroad in Oregon. Leaving Sacramento at 2.20 P. M., Redding is reached at midnight of the same day. In half an hour the stage ride begins, and lasts until about four o'clock the next afternoon, when Strawberry Valley is reached, about two hundred and forty-five miles from Sacramento — time twenty-five hours. By this method of travel, much of the upper Sacramento Valley, and of the foot-hill region north of Redding, is lost to observation, either going or returning. The night ride on the stage, over a rough road, especially in the late summer when the dust is thick, is very uncomfortable and wearisome; yet it has a certain strange interest. The large head and side lights to the stage, *alias* "mud-wagon," cast weird reflections on the deep cuts in the rocky hillsides, and on the ranks of gray-trunked oaks or dusty thickets of underbrush. At the stations, placed at intervals of twelve miles, sleepy hostlers come out with fresh relays of horses, and their half-unwilling talk with the drivers reveals queer glimpses of lonely wayside life, with its paucity of incident and topic. Here and there distant hill-sides are in a lurid blaze, — the effect of some careless camper's fire, which is spreading destruction among the noblest coniferous woods. Sometimes the stage will dart rapidly through a bit of burning forest, the ground beneath the flaming tree-trunks strewn with ashes and beds of red coals, the air heated and filled with suffocating smoke, which has a resinous odor. Three times the stage is ferried across the Pit and McCloud rivers — the main branches of the upper Sacramento, flowing to it on the east, from the northeasterly slopes of Shasta, as the Sacramento itself flows from the southwesterly flank of the same peak — cold, snow-fed streams, all three, which convey to the warm valley nearly all the chill of their origin; clear and rapid, too, the resort of myriads of salmon, which seek them from the sea in the breeding season, and the constant home of several species of trout. The foot-hill country along the Sacramento contains a few mining camps, as gold is still scantily extracted from the river bars, the ravines, and slopes. Granite gives way to slate more or less veined with quartz, and the drift revealed in the river-bed or bank is largely made up of granite, slate, and quartz, mixed at last with boulders and cobbles of trachyte and lava.

As day dawns, the foot-hills, with their several species of oak, — smaller than those in the valley, — of ceanothus, syringa, manzanita, and poison-oak, have given place to long, high, straight ridges, clothed thick with pine, and fir, and spruce. These ridges, composed of metamorphic and volcanic rocks, form a deep, broad cañon, unlike the cañons of the Sierra to the southward in this, that the river is still clear and unobstructed by mining wash, that its banks have some level space on either side, and are not divested of their beautiful vegetation, including groves of conifers, which spread down from the ridges, mixed with dark-limbed, slender, and graceful oaks. As the mining operations along the upper Sacramento are very small, and confined to the primitive methods of cradling and sluicing, no hydraulic diggings having been found, the stream retains its primitive character, and for the greater part of its length its banks are virgin. The contrast it presents to eyes accustomed to the choked and muddy streams of the deep gravel region southward, whose original banks and bars have been

buried fifty to a hundred feet in mining *débris*, and whose higher banks have been stripped of timber, is delightful. The road follows along the steep side of the ridge on the west of the river, sometimes rising several hundred feet above the stream, then plunging down to its very channel, leaving and returning to it in picturesque coquettishness. The river itself is an almost constant rapid. Having a softer material than the granite-bedded Sierra streams to cut through, it has worn its channel low down on a nearly uniform grade, and nowhere on its course, from the foot of Shasta to the plain, has it any of the falls and cascades which characterize the Sierra streams. It has a beauty all its own, however. In a succession of riffles, whose foam is tinged with blue or tea-green, it dances and sparkles and sings over its clean bed of boulders, over exposed ledges of bedrock, over bars of gray gravel. At intervals, masses of basalt-like rock rise in columnar forms or make a terrace of many-sided slabs, at the edge of the transparent current. For fifty miles the water is fringed with rich masses of very large, round, and scalloped leaves, slightly drooping from a centre stalk, big and shapely enough for parasols. These growths, a species of saxifrage, along exposures of volcanic rock that form ledges in the water or rise in cliffs above, characterize this stream to within fifteen miles of its source. Ascending its course, the ridges rise higher and higher, until those immediately hemming it in, scarcely half a mile apart, reach an elevation of two thousand feet above its level, their thickly wooded flanks plunging down very abruptly, and their straight-drawn summits bristling with arrow-headed conifers, through which, and through their hazy or smoky shades, the sunlight breaks in radiant bars, filling the whole cañon with a mellow glory. Always the rippling laugh and song of the rapid river, foaming between its green rows of parasols, with their twin rows of reflections where the water is still; always those straight, high ridges, with their terebinthine woods and floods of broken beams. Watching the river, we can often see the dark-backed salmon pushing up against the riffles, resolute to obey the instinct that reminds them in ocean depths of the cold, fresh stream in the heart of far mountains. The trout feed on their spawn, and with that as a bait can be caught with hook and line in great numbers. Bailey, of the Lower Soda Springs, told the writer that he caught in June, July, and August, 1873, three thousand one hundred and eighty-two trout, baiting with salmon eggs. And these upper Sacramento trout are beautifully speckled, with bright silver bellies, weighing commonly from half a pound to two pounds, and often more, and having a rich pink flesh.

From time immemorial the upper Sacramento and its tributaries, the Pit and McCloud, which closely resemble it, have been the favorite fishing resorts of the Indian tribes once so large and numerous in this region. Here they gathered in multitudes to spear the salmon and hold protracted festivals, of which fish-bakes, primitive gambling games, and dancing, were the leading diversions. These gatherings, though in sadly diminished numbers, still occur in the height of the summer fishing season, and at intervals along the Sacramento may be seen the conical bark huts laid up by the Indians, occasionally still tenanted by picturesque but filthy groups; while far into the stream, over deep pools, project the poles, supported on crotches, upon which the red man stands and hurls his spear — his nude, shapely form suggesting the idea of a bronze image, as, erect and still, with eye intent and arm uplifted, he poises his weapon for a throw. It is not strange the poor savages resented the intrusion of the whites upon these picturesque and productive rivers — an intrusion accompanied by much brutality and violence, compared with which the retaliatory acts of the Indians lose half their atrocity. It may possibly have been an impulse of romantic sympathy, as well as mere recklessness, which led Joaquin Miller, in his uncurbed, wayward youth, to consort a while with the Shastas. Following in his footsteps through this region, one discovers the source of much of his best poetry. On the Sacramento, the Pit, and the McCloud, he made the studies for those wild, fresh landscapes which live in his poems. Among these lofty ridges and loftier peaks, in the very shadow of Shasta, he found all his best imagery, and conceived his ideal brown beauties. Here was inspired and fed that deep fondness for wilderness life which is the prevailing characteristic of his muse. Whatever the irregularities of his career, it made him the first original poet of the western wilds. Old settlers through the upper Sacramento country have many stories to tell of him, and some are not more flattering than he would like them; but those who knew him best agree in testifying that he was a dreamy, imaginative young fellow, who loved to muse idly by river side and on the mountain top, and who, amid all the savagery and looseness which he shared, had a soul in constant sympathy with the beautiful in nature.

But to return to our journey. Following the upper Sacramento, the view of Shasta which can be had from the big valley is quite lost. Intervening mountains near the eye shut it off. One looks constantly forward in hope that these will open and reveal the supreme height. Rising from every plunge to the river to some point commanding a larger view, we look and look in vain, until within fifteen miles of the end of our wearisome staging. Then we see, first, — from a slight elevation of the road overlooking an ox-bow bend of the river, which incloses a level bar overgrown with conifers, — an abrupt and jagged ridge of bare granite, thrust up through the slate and overlying lava of the surrounding country to an elevation of two thousand five hundred feet above the valley. This ridge is a spur of the Trinity Mountains, putting in from the western side, and terminating in a peak called Castle Rock, whose extremely narrow and sharply serrated crest, of an ashen-gray color, presents the appearance of spires, pinnacles, and domes, whose sides are nearly perpendicular. The lower slopes of this beautiful ridge are covered with heavy forests of fir. — It reminds one of the Yosemite cliffs, and is probably the most beautiful uplift of granite outside of that wonderful valley. When the atmosphere is clear and the sun is in the eastern heaven, the bare rock is exposed in all its hard anatomy and native coldness of tint. But when the sun declines toward the west, the gray granite crags become violet, deepening with evening into purple, while a soft lithographic shading subdues their ruggedness and hides the detail of their sculpture. As the sun goes down behind them, the brilliant purple and crimson haze which enwraps the peak and fills forest and valley with glory, makes the scene indescribably fine. A daring engineer of the Oregon Railway climbed the tallest of the splintered rocks comprising this peak, at some personal risk. Hunters have pursued the deer to the base of the highest crag, and on one occasion a hard-pressed buck sprung over a precipice and was dashed to death below. The Indian women used to climb nearly to the top to gather the manzanita berries which grow on the sloping *débris*, until one was caught in a slide and killed by the rocks striking her head from above, with which accident they are said to connect a superstitious dread. Like the Aryans in their native seats, and their cultured Greek descendants, these simple aborigines people high mountains with supernatural beings, who are thought to be jealous of the sanctity of their retreats. On the farther side of Castle Rock is a little lake, above whose deep and still waters rise the granite cliffs with fine effect.

Continuing up the Sacramento, whose channel has now reached an elevation of about two thousand three hundred feet, we reach a group of chalybeate springs, containing chloride of soda in the largest proportion, and heavily charged with carbonic acid. The finest of these springs, eight miles from Strawberry Valley, known as Fry's Soda Springs, had formed a large mound of soda, silica, and iron before it was welled and covered for the use of visitors resorting to it regularly. Close by flows the swift, clear current of the Sacramento. A swarded peach-orchard, with its bright grass, the light foliage of its trees and their burdens of blossom or fruit, contrasts prettily with the sombre color and monotonous forms of the coniferous woods adjoining. Immediately behind the orchard rises a very straight and steep mountain ridge, quite two thousand feet above the valley, — an immense wall of forest, so precipitous that the growth of tall timber on its flank is a wonder. This ridge is a grand object in the afternoon, when the declining sun shoots his rays in long lines through its woods, turning smoke or haze into a veil of softened glory. It is while descending an incline toward the Soda Springs that the first glimpse of Shasta is caught, looming far above such a line of timbered ridges as that described, a cone bare of vegetation, of a pinky ash color where the snow has melted, ethereally soft in the hazy or smoky perspective of summer, but earlier in the season sharply relieved against a clear sky, with all its sculpture revealed, and its crown entirely white with snow. The sight of this great peak, so long sought, at so much labor, begets a sudden oblivion to dust and fatigue. The spirits are elated with a new sensation, and it is with a sigh of regret that we see the stage plunge into a dense wood, which shuts off the wonderful vision as suddenly as it appeared. For eight miles beyond the Soda Springs the road makes up a tedious ascent — part of the old lava flow of Shasta, rough and dusty; yet it should not be tedious to the lover of nature to ride through such magnificent groves of pine and fir as clothe it, wherein the sugar pine reappears after a long absence, its massive trunk frequently six feet through, and its picturesque spread of boughs, with their long cones at the ends, rising to a height of two hundred and fifty feet. In this last eight miles an ascent of about one thousand two hundred feet is made, and we reach at last Strawberry Valley and the welcome house of Sisson, weary enough, but not too weary to stare delightedly at Shasta, now in full and plain view before us.

Strawberry Valley, or Flat, as it is called by some, is the first opening into a series of small, elevated valleys which stretch about the base of the peak, extending on its western side through Siskiyou County, and including Shasta, Cottonwood, and Scott valleys to the north. Strawberry embraces an area of only a few miles, broken by encroaching belts of conifers which divide it into several parts, and bounded on the west by the lofty Scott Mountain, — a range whose crest rises at least five thousand feet above the valley, and is spotted with snow through the whole year. The northern limit of the valley is Black Butte, the highest of a large number of inferior volcanic cones dotting the plateau northwest of Shasta. From the beautiful regularity of its outline, this sugar-loaf mass of trachytic rock was named Cone Mountain by the Geological Survey; but the local and popular name is that given above, and was suggested by the dark color of the peak, which is exaggerated by contrast with the bright verdure of Strawberry Valley, and with the pallid tints of the grand mountain adjoining. Black Butte has an elevation of more than three thousand feet above the plain at its base, which makes it over six thousand five hundred feet above the sea. Away from the belittling bulk of Shasta, it would be a very imposing peak, and even where it is, by reason of its sharp and sudden uplift, and its isolated position, it is a prominent and picturesque object. Strawberry Valley derives its name from the abundant growth of wild strawberries over its surface. This delicious fruit can be picked, though in small quantities then, as late as September. A large circular area, formerly a marsh, fronting Sisson's house, and extending to the timbered base of Shasta, has been drained by the settlers — chiefly by Sisson himself — and cultivated to timothy. By means of irrigating ditches, this meadow is kept beautifully green through the whole summer and autumn, when other valleys are brown and parched. The small creeks and brooks which flow together here from Shasta and Scott mountains, forming the main Sacramento, meander through the timbered or open spaces of the valley, until they reach a common outlet into the cañon at Soda Springs. Looking from the porch of Sisson's house, with its pleasant frontage of grass-plat and flower-beds, across the timothy meadow, one sees the noble bulk of Shasta, only twelve miles off in a direct line, rising grandly above the belts of pine and fir that encircle its base. As the valley is only three thousand five hundred and sixty-seven feet above the sea, and the highest peak of Shasta is fourteen thousand four hundred and forty-three feet, it follows that the eye takes in at one glance an uplift of ten thousand eight hundred and seventy-eight feet. Seen from this place, it is a double-pointed peak, with a considerable space between the two summits, the long, sweeping line of its sides having an angle above the timber of twenty-seven to thirty-six degrees, and thence sloping down in more gradual curves, which finally melt into the valley.

Isolated by the valleys around its base from the ridges of the Sierra Nevada and the Coast Range, which in this region are conterminous, if not quite intermixed, and showing so much of its real elevation, Mount Shasta has the finest exposure of all the lofty summits in California. Indeed, there are few mountains anywhere in the world which stand so apart, and are seen to such great advantage. Mount Whitney, in southern California, — its superior in height by five hundred or six hundred feet, and its only proved superior in the United States, outside of Alaska, — is but one of a number of companion peaks, of little inferior height, rising a few thousand feet above the general elevation of a long crest-line, accessible by a quite gradual approach on horseback. The peaks about the railroad summit, having an elevation of from nine thousand to ten thousand feet, are reached by an ascent, on the railroad or wagon-road grades (which go within three thousand or four thousand feet of their tops), not less than one hundred miles long. But arrived at the base of Shasta, you are only three thousand five hundred and sixty-seven feet above the sea, and make the remaining elevation of nearly eleven thousand feet to the top, on horseback and afoot, in the short distance of fourteen or fifteen miles. Standing out so boldly, Shasta is a conspicuous landmark over an area several hundred miles in extent, and the view of it from any of the valleys at its foot is alone ample reward for the long journey necessary to obtain it. The study of it from Strawberry Valley is a constant source of pleasure, for many days in succession, from the early morning, when it is cold and austere, until the evening, when it is warm and ruddy with a delicious Alpine glow, lasting forty minutes after the valley is in cool shadow. In the clearest atmosphere, and close as it is, the twin cones of its summit look soft and smooth, as if clad with soil, where they are not covered or streaked with snow. Innocent and inviting as are those slopes, except for the steep angle of their inclination, we know they are rough piles of broken rocks, of toppling slabs, and sharp volcanic clinkers. But how lovely they look! How delicious in their prevalent tint of pinkish drab, streaked with the red of lava edges and the white of frozen snow, and relieved so high up against the blue sky; while low down is the abruptly terminating line of dark green firs and pines, sloping to the bright grassy meadow, at the foot of of all. In some lights, and especially when the atmosphere is hazy, the peak above the timber-line is a delicate *mauve* color; and it is then as airy and wonderful as the dome of Aladdin's genii-built palace, insubstantial almost as the fabric of a vision.

This description applies only to the summer aspect of Shasta, for from November or December until June or July, the perfectly clear atmosphere shows a distinct and massive cone of snow, glittering in the sun or veiled only in clouds. The amount and duration of the snow depend upon the character of the winter. If that is mild, the snow will not fall so deep nor last so long on the lower slopes as in ordinary seasons. But there is always more snow on the higher portions of the mountain than appears from the foot, especially from the valley on the southwestern side, where the influence of the sun is greatest. Depressions invisible from below will be found, on reaching them, to be wide-stretching fields of frozen snow and ice; and the northern slopes, equally with the loftiest points on top, cannot be reached from Strawberry Valley at the latest date in summer, after the mildest winter, without crossing such fields.

The winter climate of the valley is mild and equable. The snow-fall is neither deep nor lasting, and the thermometer seldom drops below the freezing point. There is not much increase in the volume of the streams, and the temperature of their water is hardly changed from that of summer, since at all seasons it flows directly from icy sources. While the winter is so bland below, on that lofty peak above it is arctic in severity, and terrific storms can be seen raging there when the valley may be comparatively exempt. Thunder and lightning are rare phenomena, usually, in California; but the great volcanic mass of Shasta acts like a magnet, and the electric storms about it are sometimes awful. The subtile fluid fuses and drills the rocky peaks at the summit, leaving large holes in the outcrop which are glazed with a green vitreous mineral, not unlike obsidian; convex blisters of the same substance adhering to the surface of the rock, and shivering to atoms when one tries to remove them. The destruction of trees by some of these electrical bursts is very great. Yellow or sugar pines, four or five feet through and two hundred feet high, will be literally torn to pieces and scattered over a wide area. One yellow pine of nearly this size, as the writer can testify, growing in a meadow near Sisson's, was torn as if by an explosion of giant powder, much of it having been thrown high up, black streaks being left along the lines of cleavage in the trunk, and the innumerable fragments of trunk and branches scattered over an area of about seven acres, disposed on the ground in rays, like the spokes of a wheel. Trees shivered by lightning, and tall splintered trunks, are frequently seen in the forests of the valley and on the lower flanks of the mountain. Even in the summer, severe wind-storms, accompanied by thunder and lightning, sometimes occur, and parties making the ascent in clear weather have been overtaken near the summit by sudden squalls, which drove them back and caused them much suffering.

The best time to make the ascent of Shasta is in July. In this month the atmosphere is still perfectly clear, and the snow is sufficiently melted to afford good camping ground at the point where the foot-climb begins. Later in the summer, the view from the top is apt to be obscured by haze or smoke; indeed, as late as September or October, before there have been any rains, the smoke from forest-fires (which were raging last year, at intervals from Redding to Yreka, a distance of one hundred and ten miles) will be apt to hide the lower country completely, inflicting a severe disappointment on the tourist. A few persons go up every summer of late years, including an occasional woman. Most of the parties making the ascent have the guidance of J. H. Sisson, whose knowledge of the country, and of its wild inhabitants, which he imparts in a pleasant manner, contributes much to the interest of the trip. He has lived about Shasta for sixteen years, is a hunter of skill and experience, and what is more rare, an earnest lover of the beautiful in nature. Under the average height of men, and weighing only one hundred and thirty pounds, he is lithe and strong, has great powers of endurance, and much courage. Educated in New York State to go through Hamilton College, a wild instinct took him west before he could enter that institution, and he found the career he loved best at the foot of Shasta, where he has made a pleasant home for his family, and is planning sagacious schemes of improvement, in anticipation of the day when the railroad shall bring hundreds of tourists every summer to the spot that he believes to be the loveliest in America. And, indeed, when the railroad shall have made the Shasta region easily accessible, it will be the finest resort, next to Yosemite, in the Pacific States, for mere scenical enjoyment, and for hunting and fishing far superior to the Yosemite, if not to any other portion of California. Deer are very plentiful in the mountains, and even in the valley thickets and woods.

Before the failure of his sight, Sisson killed from sixty to eighty a season, with his single rifle. The brown and grizzly bear, quail, and grouse are also plentiful. All the rivers are stocked with splendid trout; the McCloud River — easily reached from this point by wagon-road — containing a rare species, called the Dolly Varden, from its large, red spots, known to the Indians as the *Wye-dul-dicket*, and found in no other stream in California, and nowhere out of the State, except possibly in Oregon. This is believed to be the same fish described in some of the railroad reports as *Salmo spectabulis*. Besides the true brook or river trout, the Sacramento and McCloud contain the large salmon trout, and in the season — at its height in July — are filled with salmon. Castle Lake is one of the best fly-fishing places in the State. As this whole northern region is wild and little explored, there being few settlers apart from the stage-stations along the one road running between Redding and Yreka, game has not been thinned out or scared away, and there is an opportunity for some original exploration. The few Indians remaining are mostly domesticated, and none are troublesome. It is more nearly a virgin country than any in California, except the extreme southeastern Sierra, which is accessible only by a tedious journey of many days, off the line of railway communication.

THE BIRTH OF BEAUTY.

An old volcano, sealed in ice and snow,
   Looks from its airy height supreme
On lesser peaks that dwindle small below;
   On valleys hazy in the beam
Of summer suns; on distant lakes that flash
   Their starry rays in greenwood dense;
On cañons where blue rapids leap and dash,
   And mosses cling to cliffs immense.

Here on this height sublime combustion dire
   Once blazed and thundered, pouring down
Resistless cataracts of rocky fire,
   That from the cloven mountain's crown,
Around its flanks in every gaping rift,
   O'er meads that girdled green its base,
Spread out a deep, entombing drift,
   A tongue of ruin to efface.

In throes of terror Nature brings about
    What gives to man the most delight;
No scene of peaceful beauty comes without
    Such birth, as day succeeds to night.
A mountain gem of pearly ray serene,
    Our old volcano shows afar;
Fills all the panting soul with pleasure keen,
    And draws it heavenward like a star.

ASCENT OF MOUNT SHASTA.

MOUNTING horses accustomed to the trail, and taking along an extra animal, packed with blankets and provisions, our little party — consisting of the writer and his wife, Sisson the guide, and one of his *employés* — leave Sisson's house in Strawberry Valley at nine o'clock in the morning, bound for the top of Mount Shasta. It is a warm September day, and the lower atmosphere is hazy and pungently odorous with the smoke of burning forests. We follow the stage-road a short distance northward, the Black Butte facing us, and then turn into the woods to the right, making directly for the peak. For two or three miles the trail, which we have to pursue in single file through tall thickets, leads across level ground, shaded by a noble forest of pine, fir, cedar, and spruce, differing little from the same growths at about the same elevation in all parts of the Sierra Nevada, except that the trees are more openly disposed, in park-like groves, and have little of the bright yellow moss on their trunks which is characteristic of the Sierra forests within the line of deep winter snow. The sugar pine remains the grandest tree, but the firs and yellow pines are also very straight, tall, and handsome. The underbrush consists of the wild rose (growing here four to eight feet high), the ceanothus, the chestnut-like chincapin, a bright-leaved, fragrant laurel (locally known as the spice-bush), and more rarely the manzanita. There are also large patches of huckleberries. These thickets are often so dense that it would be hard work to follow the slight trail through them on foot; and, even on horseback, one must watch against entangling his stirrups. Hundreds of species of herbaceous plants occur, and nearly all the shrubs and plants are bloomers. When the rose thickets are in blossom, the air is delicious with their fragrance, and the honey-bee — which has become wild in these woods, as elsewhere in the Sierra — finds great stores of food. Late in the summer the balm of Gilead, which grows along the streams, distills from its leaves a sugary secretion, called honey-dew, on which the bees also feed. One swarm of bees in the valley, which was hived about the first of June, made from that time until September fifteenth, — say in three months and a half, — no less than one hundred pounds of fine honey. It is pleasant to note the absence of the poison-oak, which nowhere in California flourishes within the snow-belt, giving out all along the Sierra at an elevation of between three thousand and four thousand feet. A little to the left of the trail, as we cross the valley toward the peak, at the foot of a ridge about one thousand five hundred feet high, which is one of the lower spurs of Shasta, leaps suddenly out of the earth a foaming torrent, clear and icy cold, whose two streams at once unite and form a good-sized creek. This is the source of the main Sacramento. To see the two mouths of its exit, it is necessary to push aside a tangled undergrowth, and to bend low. Between these vents is a large chalybeate spring, which seems to have a different origin, and stains the earth between the parted, snowy waters a rusty red. There is a remarkable richness in the flora of this locality, embracing, among the bushes and small trees, species of the willow, alder, cornus (resembling the eastern dog-wood), birch, hazel, elder, black oak, yew, maple (*Acer circinatum*, probably), wild rose, chincapin, choke-cherry, black raspberry, gooseberry; and, among the smaller growths, the snow-ball, strawberry, pennyroyal, besides several vines and small herbaceous plants, ferns, mosses, and water plants. The springs that feed all this vegetation are undoubtedly the outpouring of a subterranean stream, originating in the melting snow and ice of Shasta, and drained through fissures and caverns of volcanic rock. One of the characteristics of this mountain is the disappearance of most of the torrents that have birth near its summit, through the broken rock and porous *débris* of its slopes above the timber line; and as it is well known that there are cavernous passages in the lava covering all the lower flanks and base of the mountain, nothing is more probable than that the lost streams of the peak reappear in the enormous springs of the valley. Wild animals of all kinds, including the bear and deer, at different seasons come to these springs to drink, and are especially fond of the salty water of the chalybeate spring. Riding through the forest on the lower flank of the mountain, which begins to rise from near this point, we met several deer, both going and returning, and higher up twice crossed fresh bear-tracks, and saw the recent wallow of his plantigrade lordship. There is a peculiar charm in following the trail of the various wild creatures of the Sierra woods, or catching glimpses of them in their privacy. Nothing is more fresh and graceful than the bounding movement of the deer, especially. At this season the does and fawns are seen alone, the antlered bucks having retired to more elevated places. The social twitter or anxious call of quails, hid in the thicket or trooping across a rocky open, is almost the only sound that mixes with the soughing of the trees, save the occasional heavy whirr of startled grouse, as they make a short flight for a place of concealment.

As we rise above the valley, at first by a gentle ascent, the character of the forest changes. The pines are less frequent, the firs are more so, and the undergrowth is less thick and varied. The ten or twelve species of conifers are reduced at last to three or four, — yellow pine, Douglas spruce, and a large fir. The surface becomes rough with broken masses of basalt and other lava rocks, part of the outflow of the slumbering volcanoes above. An unusually rugged field of this material, where vegetation is nearly exhausted, and where the horses bruise their pasterns at every step, is called by Sisson "The Devil's Garden." At an elevation of about seven thousand feet the pines give out entirely, and we go through a belt of silver-leaf firs (*Picea nobilis*), a very symmetrical, beautiful tree, with a juicy, greenish-tinted bark, foliage of a faint tea-green color at the tips, almost silvery in certain lights; the trunk small in diameter, but straight and taper as a mast, and reaching a height of one hundred and fifty feet. These handsome firs scent the air, while they shut out the rays of the sun and give the sky a darker color as seen through their dense capitals. The beauty of the trees on the lower flanks of Shasta has become known in Europe, where their seeds are in demand. Sisson has orders for forty to sixty pounds of coniferous seeds yearly, from Germany alone. As the small cones of the silver-fir grow at the very top of the tree, he has to climb one hundred and fifty feet to get the choicest. From the lower boughs of many of these trees hang long streamers of black moss, curiously like coarse human hair, and calling up fancies of Absalom caught by his tresses. On the upper edge of this belt of silver-firs we come upon the path of the avalanche. Vast snow-slides have mowed wide and long swaths through the timber, strewing the earth with broken trunks and branches, which are partly buried in ash-like *débris*. The boundary of these slides is often marked by a bright, little grove of young firs, more delicate in color than the adjoining forest. In this region of avalanches, also, the *Pinus flexilis* — last tree to maintain life in the upper Sierra — begins to appear as a shrub, becomes a small tree as the firs give out, and expires as a shrub again at the last limit of vegetation, save moss or lichen. The forest growths cease quite abruptly on Shasta at a height of about eight thousand feet, though the *Pinus flexilis* maintains a scattered and precarious life for a thousand feet higher. This pine, with its light-gray bark wrapping the twisted and gnarled trunk as tightly as a skin, with its contorted and depressed limbs bearing brush-like bunches of bright green needles, is a very characteristic production of great elevations in California. It roots itself in the very rock, and has the aspect of strenuous struggle with unfriendly elements. Its flattened top is often so compacted by the deep snows that a man can stand upon it, and when the bushes grow thickly together, he can almost walk from one to another. Where patches of it have died at last in sheer despair, it still stands in obstinate strength, white and weird, a stubborn ghost of a dwarf tree.

On a bold bluff overlooking a deep gorge on either side, and composed of red lava, broken and weathered, but still lying in the place of its flow, we reach at last a camping place, above the line of vegetation, as of perpetual snow, and between nine thousand and ten thousand feet above the sea. It is nearly four o'clock, and we have been almost seven hours making twelve miles of distance, and something over six thousand feet of elevation. Our horses are tired and lame, and we are glad enough to give them rest. In one of the gorges, a few hundred feet below our camp, there is a feeble growth of bunch grass, at the edge of a field of frozen snow, which they are led to pasture upon, after short rations of barley and a drink of snow-water. It was curious to see one or two of the animals tasting the snow, as they were driven across it to the drinking-pool formed by its melting during that day. Gathering branches of the dead *Pinus flexilis*, we made a fire against a mass of lava, spread our blankets within circular walls of lava rocks, piled up by previous climbers as a shelter against the cold winds, and prepared for supper. Within a cranny of one of these open bed-chambers we found vessels of tin and iron, for boiling and frying, stored there by the provident Sisson, who soon got ready a grateful meal. After tethering the horses near by, we were ready for night, intending to eat breakfast, and start on our foot-climb up the peak by day-break.

The scene about us was wild and desolate in the extreme. Our camping ground, as before stated, was a bluff bench of red lava and clinker, above the general surface of which were heaped at intervals huge detached masses of the same material, that had fallen down from above or become detached in place. The outer edge of this bench commanded a view of the whole southwestern slope of the mountain down to

Strawberry and Shasta valleys, over six thousand feet below; across the valleys to Scott Mountain, overlooking the Black Butte, which, from this height, was diminished to a small mound; and thence southerly to the cañon of the main Sacramento, bounded by long and hazy ridges, and filled with smoke from forest fires, which obscured an otherwise magnificent view. The flank of Shasta itself was marked by trough-like grooves, evidently cut by the melting and sliding snow; the timber growing to the edges of these grooves and then giving suddenly out, except where it came in as an unbroken, solid belt lower down. A large meadow-like plain, four thousand feet below, we knew to be a thicket of tangled and thorny bushes, threaded only by deer. As the sun sank toward the crest of Scott Mountain, through dense strata of smoke, it became a blood-red globe, quite shorn of its beams, and more or less elongated, and could be looked at steadily. It was very strange to see this red ball dropping through one band of smoke after another, for the strata were of unequal density and width, and the sun seemed to be sinking behind bars that made it visible only occasionally and partly. Looking backward to Shasta, its highest peak was in clear sky, and rosy bright, — a massive cone of lava-blocks and snow. To the right and left were deep gorges putting down from the peak, their basins filled with snow and ice, their slopes partly covered with long, narrow bands of snow which led up to the top at a very steep angle. Numerous torrents pouring down the upper slopes gave forth a subdued roar, varied by the dull rumble of the rocky masses they detached, and which seemed, by the sound, to be constantly moving, although we could not see them. The red lava bed on which we stood extended for a mile or more, at a slight inclination, to the very base of the peaks, which it surrounded like a garment that had been pushed down, leaving the two cones of the summit standing clear above, of another color, their outlines drawn sharply against the sky, — preeminent, lonely heights, their tops as far above our exalted station as Mount Diablo or St. Helena above the sea, — literally, Pelion on Ossa. For we can now see plainly the true shape of this volcanic mountain. Its apex is divided into two craters. The one at the left hand, the lower of the two, is shaped like a sugar-loaf, with the top cut off; yet above the circular rim of this flat top rises a small pyramid, giving the whole mass a very peculiar appearance. The right hand and higher peak is less regular and formal in shape. Its northerly slope comes down to join the left hand cone in a sharp, clean line, the depression between being filled with a broad field of snow; but the southerly slope has a much reduced inclination, running to the timber line far below, and its knife-blade edge, composed of volcanic conglomerate, is broken into the most fantastic shapes, suggesting castellated structures at times, but oftener the forms of gnomes and demons. The Indians imagine these weird shapes to be, indeed, a kind of mountain sprites, which they call *appetunes*, and which appear in watchful and observant attitudes, as if on guard against mortal intrusion. The face of this peak, between the outlines, is a steep bluff, depressed below the wall-like upper edge of bright red breccia, and scarcely half covered at this season with long bands of snow. The summit has several sharp points, which rise above the basin of an ice-filled crater, invisible from below, as is the basin of the left hand crater. The lower peak — called distinctively "Crater Peak" — is a uniform chocolate-drab in color, viewed closely; while the higher point — called "The Main Peak" — diversifies this color with its bluff and ragged edges of red breccia, with a band of black rock and beds of ashy *débris*. Late in the summer the snow is quite gone from the surface of Crater Peak upon the steep southern side, remaining always at the top, however, and in the depression between it and the other peak. The southern face of the Main Peak is never free from snow. As measured by the State Geological Survey, the outline of Crater Peak has an inclination of 36°; that of the Main Peak has an inclination of 27° to 28° on the shorter, and of 30° to 31° on the longer side. As we contemplate these outlines from below, the task of climbing either of them seems formidable enough, and it is certain that portions of the slope to be passed over are steeper than the measured outline. According to Professor Whitney's "Report," published in 1865, the Crater Peak had then never been ascended, and was "believed by many to be quite inaccessible." Its sides, he adds, "appear to be covered with loose volcanic materials, probably ashes, lying at the highest angle possible without sliding down." The steepness of this cone was not exaggerated, but it has since been frequently climbed, and has latterly been included on the route to the Main Peak by a few of the strongest and most resolute climbers. In 1871, Clarence King's party, which spent six weeks on and about the mountain, scaled up this side cone with instruments, including the photographic apparatus of Watkins. If the slopes were really formed of ashes, or other fine material, they could, indeed, hardly be climbed, as they would offer no secure footing at such a steep angle; but they are covered with angular blocks of trachyte, sometimes very large, formed by the breaking down of the crater walls above, and affording a footing in the steepest places. From our camp, these rough slopes looked smooth enough to be ash-beds, and the distance to the top, though several miles, and involving an ascent between three thousand and four thousand feet in perpendicular height, seemed to be very short in that clear, upper air. Nearly one third the atmosphere which men breathe was already below us, and the exertion of bringing wood and water to camp and spreading our blankets for the night made us pant. Thus the stratum of atmosphere above was thin and clear. The early stars as they came out were unusually large and lustrous, and later, when twilight was quite gone, the heavens seemed as populous with bright points, and as luminous, as in southern latitudes. After nightfall, the temperature of the air was at the freezing-point, and as the snow ceased to melt, the roar of the torrents stopped, and no sound broke the awful solitude of the mountain after we took to our blankets, except the occasional stamping of the horses on the clinking lava.

It was not easy to sleep in such a place, with that brilliant heaven above, and the massive front of the peak projected like a shadow against the eastern sky, save where its long streaks of snow gave it a ghostly pallor. We often woke, and gazed long at the glorious vision overhead, or on the severe outlines of the peak. At last Sisson arose, declared day was about to break, and began making a fire. It seemed impossible the night was so near gone; yet there in the east, right over the shoulder of the mountain, was a pale silvery glow that appeared to herald morning. It brightened, but with a brightness like that of the moon, and just then the planet Venus, large and lambent, — "like a rich jewel in an Ethiop's ear." — rose above the fantastic outline of the mountain to the right. Attributing his mistake to the singular purity of the air at this altitude, Sisson was fain to seek another nap. It was not long until daybreak, however, and we had an early meal, shivering until warmed by the hot tea. This dispatched, we began the ascent of Crater Peak, wearing our thick woolen clothes, and carrying iron-shod and spiked alpenstocks, a tin flask of cold tea, and some food, a man remaining behind to care for the horses. Reaching a more elevated part of the red lava field, we could see the first light of the sun on the lofty crest of Scott Mountain in the west, Shasta before us being still in cold gray, its enormous cone preventing the light from falling on its own westerly side, and casting a sharply defined pyramid-shaped shadow thirty miles long over the valleys at its base and the mountain range beyond, all outside of this dark purple shadow being in sunlight as we looked wonderingly below. We met the first direct beams of the sun as we reached the foot of Crater Peak, and now began to realize the rocky roughness of its slopes. Going up these was like climbing very steep stone stairs, except that the steps were uneven and often unsteady, — one rock tipping on another, so that each planting of the foot had to be calculated to avoid slipping or toppling, — and that the placing of the alpenstock, which was an indispensable support, had also to be studied. Breathing became more and more difficult in the increasingly rare atmosphere, and but a few yards could be climbed without a rest. The beating of the heart was audible to each person, a pallor came over the face, and the eyes were strained in their sockets. As we looked upward from time to time, the rim of the flat top seemed no nearer. As we looked down, the large blocks we had overcome grew small, and the apparently fine *débris* ahead grew large when we reached it. The big snow-fields on either side of our camp shrunk into little patches; we could no longer distinguish the camp itself, nor the horses. The steep edge of the rounded cone on the northerly side was drawn down across the sky in one tremendous line of rock that seemed a jumping-off place into the nether air. We were insects crawling up a slanting steeple, far above the world. The view below was awful in its depth and extent, the still obscuring smoke giving it a character of mystery and indefiniteness. There seemed no bound to that blue, hazy gulf; and above, to the left as we climbed, was only the lofty sky-line of the cone, stretching up, up, up. An occasional field of fine *débris*, which slid under our weary limbs, made us glad to regain the securer blocks of trachyte. On the latter we could sit, as on benches of stone, panting, perspiring somewhat as the sun's heat was reflected from the bare, smooth rocks, but always enjoying the grand sensation that comes from being high above the world, on a narrow point of its crust. Under our feet, as we climbed, we heard constantly the gurgle and murmur of an unseen torrent, fed from the melting snow above and running deep below the thick-piled masses of rock over which we stepped. For two miles or more we climbed above the channel of this hidden stream, never once catching the slightest glimpse of the water. All around the mountain there are subterranean torrents like this, which go to form the great springs that leap into rivers at its foot, — " water, water everywhere, nor any drop to drink," except that in the flask you carry.

At last we reached the rim of the flattened cone above, but not yet the top of Crater Peak. There was a narrow snow-field to cross, lying in a depression, and then a small pyramid of broken trachyte, about five hundred feet high, capped with a portion of the original crater wall, to clamber up. It was eleven o'clock before we reached the latter point, which presented itself as a perpendicular ledge about twenty feet high, but so creviced and broken that we got easy hand and foot hold, and so pulled up to the top, where there was just about room enough for our party of three to recline. This narrow ledge is the very summit of Crater Peak, and is nearly thirteen thousand feet above the sea. We found here the small monument left by Clarence King's party two years before. We had been five or six hours toiling for this mark, experiencing much difficulty in breathing, and even nausea, from the effects of the highly rarefied air. The weather was unusually warm for the locality, and no clouds obstructed the direct rays of the sun. The climb was, therefore, more fatiguing, and respiration more difficult, than they would have been had a cold air been blowing, or had the sun been overcast. Sometimes parties who make the ascent in the same month (September) encounter bitterly cold winds and storms of snow. Thomas Magee, who described his ascent in "Scribner's Monthly," found the cold so severe that it partly froze the tea in the tin canteen at his side. But warm or cold, the view at the summit amply repays all toil and hardship. Even if the lower country be hidden in smoke, as was partly the case in our experience, the mountain itself is a grand sight and an instructive study. Standing on the pinnacle of Crater Peak, its sides are seen to descend at a steep angle all around, and one has almost a dizzy sensation on realizing the immense depth into which he could plunge by a slight effort, or tumble by a reckless step. On the north side, immediately beneath the eye, lies the old crater, — a circular cavity a mile across and a thousand feet deep,— its bottom and part of its steep outer and inner slopes covered with snow and ice. The wall of the crater is broken, as one would break out the side of a bowl for a quarter of its circumference, on the northwestern side, above Shasta Valley. The edges of this break must be one thousand five hundred feet long, and through the enormous gap thus made one looks from the cliff above clear down to the valley at the base of the mountain, nearly nine thousand feet, the angle of the view being fearfully steep. Shasta Valley is seen to be dotted with small volcanic cones, — miniatures of the Black Butte,— and beyond, along the western sky, are the Scott and Siskiyou mountains; and beyond these again, if the air were clear, we could see the straight leaden line which marks the Pacific Ocean. On the southerly side of Crater Peak its slope descends to a wide gorge one thousand two hundred or one thousand five hundred feet deep, filled with frozen snow resting on a substratum of ice, beyond which rises the Main Peak, more than one thousand five hundred feet higher than the top of Crater Peak. Its northern slope is regular and abrupt, but its crest is broken into several craggy points, chief of which are three needle-like splinters rising above a large basin and forming part of the walls of a crater; while the southerly slope runs off in a long, curving, broken line, fantastically ragged on its sky-edge of highly colored breccia. On the summit are sulphur springs, hot enough to boil eggs, and considerable deposits of sulphur — the last relics of the former tremendous volcanic activity which covered with lava all the slopes and valley bases of Mount Shasta, for more than a hundred miles around. What remains of the crater on the Main Peak is filled with ice to a great depth, and from this source, through a cleft on the northeasterly side, descends the slow moving mass of the Whitney Glacier, — a genuine river of ice, half a mile wide and perhaps seven miles long, — the true character of which was first determined by Clarence King so recently as 1871. All the northerly flanks of the mountain are largely covered with snow and ice above an elevation of eight thousand or nine thousand feet, and on that side, also, there is another deep gorge between the two peaks. Leaving our perch above the lower crater, we crawl down the ledge toward this gorge, and cross a small pond of smooth blue ice at its base. It was on this level spot that Watkins pitched his field-tent for photographic work, and when he thought he had the light all shut off, found that enough still came through the ice floor to spoil his negatives, obliging him to cover that also. The surface of this ice, as of the large snow-field adjoining, was slightly melting. But the air was sensibly cooler on this side of the mountain, and it was a relief to be walking again on a comparatively level surface.

At the right of the crater there is a long dike of crumbling siliceous and sulphurous rock, which we traced half a mile in a direction nearly east and west, resembling one of the metalliferous lodes in its structure, having side walls of trachytic rock, and being filled for a width of two or three feet with a white pasty mass, which on exposure hardens to the appearance of silicate of soda, more or less discolored with sulphur, fumes of which still came up through this curious vent, scenting the air. Here we rested for half an hour, ate our luncheon, and gathered specimens. A slight descent brought us to the rim of the crater wall, sharp as the edge of a roof, and its snowy slopes descending on either side steeper than the angle of a roof. The melting crust on this rim was just wide enough for us to walk in single file, covering our eyes with gauze to protect them from danger of snow-blindness. The crust had been carved by alternate melting and freezing, aided by the wind, into furrows with knife-blade edges, which would make hard walking on cold days. But warm as the day was, it was interesting to observe how slightly its influence penetrated the frozen snow and ice. Even on the steep slopes of broken rock, where no snow was visible, we found that ice was spread everywhere at a slight depth below the surface; and as we laid down where this *débris* was finer than usual, it began to melt only with the heat of the body. Digging a little with the iron point of an alpenstock, we found ice where we had not before suspected its existence, and the surface-melting of these covered ice-beds was the cause of many of those hidden torrents which ceased to run and roar after night-fall.

Leaving the curving roof-line of the crater edge, and walking along the side of an abrupt incline of loose *débris* largely made up of such materials as composed the curious dike above described, we came to a projecting point where we could look up and down the northerly slope of the Main Peak, and could plainly trace the course of the Whitney Glacier for five miles. The peak on this side is three-pronged, and the glacier heads up between two of the prongs. Beginning at an angle sharper than any previously noticed, it soon assumes a gentler incline, and finally reaches the lower slope of the mountain nearly on a level, broadening at this point to its widest dimensions. The head, and all the steeper part of the glacier, present a surface of clean, marble-like *névé*, marked with numerous transverse crevasses, which open very large cavities and expose walls of blue ice. The upper side of the first crevasse, near the head of the glacier, seemed to be quite sixty feet above the lower. The difference in elevation of the crevasse walls lessened, of course, with the reduced angle of the glacier's inclination, until these openings were simply even gaps across the ice. A mile or two below the summit the surface was burdened and partly hid with lateral moraines, which lower down completely hid the ice, save where the black *débris* was parted by an occasionally wide crevasse, or a portion of it had sunk bodily into the ice, leaving a cavity filled with blue water. The morainal matter had accumulated in one place to a height, apparently, of not less than fifty feet. Owing to the mildness of the preceding winter, when comparatively little snow fell, followed by a very long season of clear, warm weather through spring and summer, the surface of the *névé* was much reduced in thickness, and the line of recent glacial cutting along the banks of volcanic material was boldly exhibited. The northerly and easterly slopes of the mountain, which are bare of timber far below the timber-line on the other side, are composed of blocks of trachyte, lava, and pumice, succeeded by an extensive outflow, lower down, of basalt. Into this material the stream flowing from the Whitney Glacier sinks, disappearing under the mass of the terminal moraine.

Beyond this glacier, easterly, is a smaller one, named variously the McCloud and Mud Creek Glacier, which was partly visible from our last point of observation. We could hear the larger ice-stream constantly cracking, and at intervals heavy detonations succeeded to this sound. We could hear, also, the roar and rumble of torrents in half a dozen different directions. But Shasta bears on its easterly flank a still greater glacier,— one not less than three or four miles wide,— which was named by its discoverer, Clarence King, the Agassiz Glacier. A trip of sixty miles around the base of the mountain is necessary to approach it, so we caught no glimpse of it. Mr. King, in his fascinating record of "Mountaineering in the Sierra Nevada," has described its appearance, and his perilous climb over it, with vivid power. One remark he makes with reference to it applies generally to the other glaciers on Shasta; it is this: "The idea of a mountain glacier, formed from Swiss or Indian views, is always of a stream of ice walled in by more or less lofty ridges. Here a great curved cover of ice flows down the conical surface of a volcano without lateral walls, a few lava pinnacles and inconspicuous piles of *débris* separating it from the next glacier." Except towards its head, the Whitney Glacier evenly fills the depression it occupies, much as the Sacramento River fills its channel on reaching the broad valley.

Apart from its isolation, the sudden uplift of nearly three fourths of its entire bulk, and its peculiar beauty of color, Mount Shasta is remarkable for being the only mountain in California whose flanks are burdened with living glaciers. The ice-field on Mount

Lyell, in the Yosemite region, which has been described as a glacier, is asserted by Whitney and King not to deserve that title; although Mr. Muir, who has given the subject close study, declares that on Mount Lyell and on several companion peaks true glaciers exist, but of feeble vitality. The taller peak of Mount Whitney, five hundred miles south of Shasta, in a latitude where the snow-line extends much above the limit in northern California and Oregon, is without a glacier, as it is also without those singular fields of rock-covered ice which exist on the upper slopes of Shasta. With the exception of this beautiful California peak, no mountains in the United States bear true glaciers but Mount Hood, in Oregon, Mount Rainier and adjacent peaks, in Washington Territory, and the Arctic peaks of Alaska, whose glaciers push quite down to the sea and send off fleets of icebergs. The grand glacier on Mount Rainier, discovered, we believe, by officers of the United States Coast Survey, has been described to the writer as rivaling, if not surpassing, anything in the Alps. Considering how easily Shasta can be reached, and with what perfect safety it can be climbed and examined, except on the larger ice-fields, it is remarkable that it is not more sought by tourists. A knowledge of glacial phenomena is now universally acknowledged to be of leading importance in the study of the earth's superficial conformation, and much could be learned in this field of inquiry on Shasta, where not alone living but the track of extinct glaciers may be profitably observed, for in every direction around the mountain exist the evidences of former glacial action.

It was with great reluctance, in the middle of the afternoon, that we left our perch overlooking the Whitney Glacier to return to camp. It was hard work to climb up the slope of sliding *débris* we had just descended from Crater Peak, and our legs trembled when we reached the icy rim of the crater and faced its blinding glare. Resting again at the very top, we gazed lingeringly at the higher peak to the left, with its cascade of *névé* and ice plunging down so precipitously for thousands of feet; at the deep crater bowl to the right, almost under our feet; at the cone-dotted, yellow, hazy valley of Shasta, seen through the broken wall of the crater over a mile and a half below; at the violet crest of the Scott Mountain range beyond, and the dark cone of Black Butte thrust up in the trough between. But for the smoke, we should have seen to the northward the whole Klamath region, with its lakes and lava-beds, where the Modocs played their miserable tragedy; should have seen the snowy peaks of the Oregon Cascade Range; should have seen to the east the desert plateau of Nevada as far as the Utah line; should have seen to the south the trough-like valley of the Sacramento nearly to the mouth of that stream, with all the bold crest-line of the Sierra Nevada range on one side, and the softer swell of the Coast Range on the other, with a strip of the Pacific Ocean near Humboldt Bay. Mr. A. Roman, who was one of a small party that climbed Shasta in April, 1856, — a most perilous season, — told the writer that the atmosphere at that time was wonderfully clear, and the view simply stupendous. He declares that he saw distinctly all the high peaks, from the Washington group on the north to the Sierra peaks around Lake Tahoe, and the Coast Range peaks about San Francisco, — a distance on a direct line of nearly eight hundred miles! Within the limits of this view the Sacramento Valley and the topography of the Sierra Nevada were, he says, revealed with wonderful distinctness. The air was as if purged and filtered, and presented only a slight gray film between the eye and the most distant objects. There seemed no limit to the vision except the convexity of the earth's surface. Probably in very clear weather the view extends for quite five hundred miles. Mr. Roman's party, and himself in particular, suffered dreadfully from the cold on the summit. He took a thermometer from his clothes to observe the temperature, and as he held it in his hand the mercury speedily dropped to 12° below zero. How much lower it would have gone he could not tell, for his stiffened fingers lost their grip, the instrument fell from his numb hand and was broken. He was snow-blind and frost-bitten on returning to Yreka, and so altered in appearance that his own brother did not know him. Sisson told us that he had been up the mountain much later in the spring, or in early summer, when the winds were so cold and strong that he had to cling to the rocks with his hands, when scaling the summit of the Main Peak, to prevent being blown off and hurled to destruction. Yet as we had this talk the air was no cooler than that of a balmy winter day at San Francisco, and our thick woolen clothes, while we exercised, were almost burdensome. Mr. John Muir, who ascended the mountain alone in November, 1874, encountered a snow-storm on the very summit, but his hardy habits protected him from injury. Waking one morning after it subsided he saw a sublime spectacle, which he thus describes: " A boundless wilderness of storm-clouds of different age and ripeness were congregated over all the landscape for thousands of square miles, colored gray and purple, and pearl and glowing white, among which I seemed to be floating, while the cone of Shasta above and the sky was tranquil and full of the sun. It seemed not so much an ocean as a land of clouds, undulating hill and dale, smooth purple plains, and silvery mountains of cumuli, range over range, nobly diversified with peaks and domes, with cool shadows between, and with here and there a wide trunk cañon, smooth and rounded as if eroded by glaciers."

Resting on the top crag of Crater Peak before descending, we observed more closely the utter absence of vegetation for thousands of feet below. After leaving the *Pinus flexilis* at our camp on the lava, where there were sparse bunches of a hardy grass, and a few plants like portulacca growing in shady crevices, an occasional lichen was all that appeared, and at the summit the lichens were no longer to be seen. On one snow-field there was a slight trace left of *Tococcus nivalis*, — the "red snow," so called. — a very low form of vegetable life, which is sometimes so abundant on this mountain as to color the foot-prints in the snow blood-red. For three or four thousand feet below, the eye took in nothing but a wreck of rocky matter, of red and black lava-flow, of gray-colored scoriaceous *débris*, except where the snow and ice covered the surface and made it even more arctic and desolate. Yet animal life was not quite absent. Lifting a piece of loose rock near the surveyor's monument, we revealed a little colony of lady-bugs, of a dark cinnamon color, with many darker spots. The tiny creatures crawled away feebly, making no effort to fly. What they could live on there we could not conjecture. A few snow-birds were twittering a thousand or two thousand feet below, and nearly up to the very crest of the Main Peak we saw a solitary California vulture wheeling slowly around. Sisson says he once found a dead squirrel on that peak, which had probably been dropped there by a bird of prey, and at another time he saw there a living mouse. The large-horned mountain sheep, apparently the same species as that found in the

Rocky Mountains, has occasionally been seen near the summit, and once an animal thought to be an ibex was observed.

Going down the rocky slope of Crater Peak, we heard again the gurgle of the hidden torrent. The descent was very tiresome, and a little hazardous to one's limbs, for a fall among the larger masses or a slide in the small *débris* might easily result in a fracture. Earlier in the year much labor is saved by sliding down on the snow. But we reached the base at last in safety, very weary, and glad to put foot again on the lava-flow that led to camp, where we arrived almost too weary to care for the red sunset through bars of clouds, which was repeated in the western sky, reminding us of the appearance of that luminary to Campbell's "last man." How sweet sleep was that night! No more deception with the morning star. Again at sunrise, however, we were off, this time mounted and bound homeward. Facing the west as we rode down the slope of the mountain, we saw once more the sharp cone of its shadow, lying far across the valleys at its foot, up the flank of Scott Mountain beyond, and across its snowy crest, the faint light trembling along its purple edges and gradually crawling into its place as the shadow of the great peak retreated. The trail down the mountain is steep and rough for horses, and very tiresome for riders. The comparative level of the forest-belt is welcome. In the black soil of the fir wood we often saw the fresh track of bears. Arrived at the only spring on the way down, we saw three deer. The graceful creatures moved off very slowly and safely, Sisson with his gun, fortunately for them, having turned into a side trail some distance back.

At the house in Strawberry Valley once more, after a journey of two and a half days, we turned to look at the grand peak with its twin cones — all its ruggedness gone, its long outlines and vast front smoothed by distance, and a sunny haze clothing it in tender beauty. Often since we have revisited it in dreams, and longed, on waking, for its restful solitude.

THE MEADOW LARK.

BERKELEY, FEBRUARY 23, 1874.

Trill, happy lark, thy brief, sweet lay,
    From out a breast as brown
As were the hills in autumn day
    Before the rains came down.

The beaming sun, the dripping showers,
    Are in thy simple notes;
Earth smiles to hear in grass and flowers,
    And bright the cloudlet floats.

On Alameda's mountain line
    The violet's tender hue,
With dappled spots of shade and shine,
    Is painted 'gainst the blue.

The meadow slopes to meet the bay,
    The gulls in flocks uprise;
And far above the waters gray
    Soars purple Tamalpais.

Beyond is ocean's wide expanse,
    Where, through the Golden Gate,
The ships with snowy canvas dance,
    Or on the breezes wait.

Fair day, bright scene! The hill, the tree,
    The poppy's running flame,
The silver cloud, the sunny sea,
    Spring's coming all proclaim.

But sweeter, dearer, far than all
    I love the liquid sound
That from the sky the lark lets falls
    Whene'er he spurns the ground.

Though all too short, his carols give
    Back to my heart once more
The thoughtless joy that used to live
    In happy days of yore.

GOLDEN GATE, FROM CONTRA COSTA HILLS

THE GEYSERS.

Yosemite, the Big Trees, and the Geysers are thought by California tourists to be the great wonder of the Golden State, next to her matchless climate and the modesty of her people. Much has been written about the marvelous gorge in the Sierra, where rivers are flung over granite precipices; and the diameter and altitude of the giant sequoia are familiar enough to the ordinary reader; but less has been said about the Geysers, although they possess features of remarkable interest. Geysers they are not, in the sense in which the word is usually understood; and the traveler who expects to see, on reaching this locality, high fountains of boiling water like those in Iceland and the Yellowstone region, will be disappointed. Yet are they richly worth the journey, as the journey itself is its own sufficient reward without any other motive than the scenery along the route. Suppose, reader, you have crossed the Sierra Nevada, breathed its exhilarating air, scented with the aromatic odor of its magnificent pines and cedars; been enraptured with the softer beauties at its base, hazy with the heat of its golden summer, or stretching far the clear perspective of its verdurous and flowery spring, and then have met on the Bay of San Francisco the cool air that blows in from the Pacific through the Golden Gate; you still have not exhausted the contrasts and pleasures of California scenery. Resting a while in the many-hilled metropolis, which sprawls over a narrow peninsula of sand and rock, resolve to go to the Geysers before you try the all-else-belittling grandeur of Yosemite. This is the route. Besides the broad Sacramento Valley, two narrow Coast Range valleys open from the bay on the north,— Sonoma and Napa,— each some forty miles long by an average width not exceeding three miles, nearly level, and bounded by high ridges of metamorphic rock of the cretaceous period, which sometimes break down into low-rolling hills that invade the plain, giving its surface a picturesque variety. Napa Valley — named from a nearly extinct tribe of aborigines — is the inner one of the two. Like its companion, it is traversed for a part of its length by a creek, navigable so far as the tide extends, which empties into the bay through a wide expanse of salt marsh. Through either valley the mountain road that leads to the Geysers may be reached. The usual route, however, is through Napa Valley. A steamboat sail of twenty-five miles from San Francisco to Vallejo begins the trip delightfully, affording a fine view of the city, — dusty, gusty, and gray on its vaporous heights; of the grimly fortified Alcatraz Island, which lies like a snag in the mouth of the harbor; of the Golden Gate, with its red brick fort on one side, its white light-house on the other, and brown or green headlands, fleets of inward or outward bound sails passing between; of Mount Tamalpais, that lifts its purple cone in tender beauty to the right, nearly two thousand six hundred feet above the sea it overlooks; of the Alameda slopes and ridges that bound the eastern shore, topped by Mount Diablo, a still higher peak; of the red rock islets in the upper bay, whitened on their tops by birds that hover or settle there; of the bare, low, mound-like hills that open as the boat approaches Mare Island Straits, that are either brown or green according to the season, but always graceful in outline, and suggesting rumpled velvet, with their slight indentations and mottled shadows; over all this varied scene a blue sky dashed with gray, which reflects its own hue in the dancing, sparkling waters of the bay and melts into hazy lilac around the hilly horizon.

Mare Island, the site of what is at present the most important navy-yard in the United States, is a long, flat body of land, very slightly elevated above the water, and on the western side of the straits. The opposite shore is hilly, its lower slopes covered with the thrifty town of Vallejo, once the capital of the State, and now the railroad and trade centre of the northern coast-valley region. Here we take the cars for Napa and Calistoga, beginning a railroad ride of forty-four miles through the Rasselas Valley of reality, whose charms surpass those of Wyoming as much as the red tints of this semi-tropical clime surpass the cold colors of the north. The trip is usually made toward evening, when the atmospheric effects are most beautiful. As the valley is filled with settlers and contains half a dozen pretty towns, its surface is marked with cultivated fields, with rich masses of green or golden grain, orchards laden with blossoms or fruit, vineyards whose cleanly kept vines shine in the sun as though they smiled over the genial harvest they are maturing. The natural features of the valley are park-like groves of oak, which grow thickest where they belt the course of the creek, and are there mixed with sycamore, alders, willows, and a plentiful undergrowth of wild vines and bushes. The spaces in the oak-openings which are not cultivated are free from underbrush, the soil bearing a native crop of wild oats and flowers, the deep

Valley Oaks

orange tint of the large California poppy (*Papavera Eschscholtzia*) being conspicuous among the latter in spring and summer. When the oat-crop is ripe, its brilliant gold colors the landscape in every direction over the valley, far up the lower slopes of the adjoining ridges, and often even to their very tops. The several varieties of evergreen oaks, with their short trunks, cauliflower-shaped masses of intensely dark green foliage, and sharp shadows, then seem like oases in the hot expanse — grateful islets of verdure in a sea of shimmering yellow light. On the rolling lands most exposed to sea winds, the oaks, contorted, dwarfed, and thorny-leaved as the holly, nestle together in groups and fit their slanting boughs to the outlines of the hills, making cool, sequestered bowers of the most inviting character. Towards the upper end of the valley the massive trunks, tall forms, and expansive foliage of the deciduous oaks, present a striking contrast to these hardy dwarfs who have to struggle for life. The willow-oak, remarkable for the pendant strips of leafage nearly touching the ground, from which it derives its name, is particularly conspicuous. One notes, too, the great rounded masses of mistletoe clinging to several varieties of oak, and the scarlet-leaved vines that sometimes cling about their trunks, rivaling in color the plumage of the woodpecker who digs his acorn-holes in the bark above. Darting through one of these noble groves, venerable with mosses, one has charming views of the mountains on either side the valley, their ravines dark with timber, their upper slopes clad with pine and fir, their northern and sea exposures luxuriant with forests of the redwood, own cousin to the *Sequoia gigantea.* The outline of the ridges is sometimes made very picturesque, not to say fantastic, by outcropping masses of metamorphic sandstone, cut into mural or battlemented shapes by the elements. When the atmosphere wraps them in its haze, and they recede into skyey blendings of all violet and purple tints, their contrast with the softening gold and green of the valley-levels is most exquisite. And when the sun sinks behind the more distant mountain masses they glow through as if molten and transparent, or no more substantial than the cloud that may be burning above them, until the sun gathers back to himself all the arrows he shot over the plain, and the slant shadows spread, mingle, and deepen into twilight.

At the head of Napa Valley stands Mount St. Helena, the culminating point of the ridges between the Bay of San Francisco and Clear Lake. It is a mass of volcanic rock four thousand three hundred and forty-three feet high; the apparently single point of its cone, like nearly all volcanic peaks, separating into two as it is approached or circled. Most of its bare bulk is visible, rising like an irregular pyramid at the end of the long valley-vista, — a grand object far and near, whether in its customary suit of gray or flashing in the splendor of its evening robe; continually shifting its color and form as it is seen close or far, on this side or that; opening its rocky breast at last to nature's softening touch of spring and brook and tree, and drawing up about its awful flanks some of the verdurous beauty of the valley. One of the best views of this mountain, on its southerly side, is that from Calistoga, where the cars leave the

Mount St. Helena.

tourist at night, and where he takes a coach for the Geysers. Calistoga is at the head of Napa Valley and the mountains here inclose a small circular plain studded with large oaks, and charged with thermal springs that send up little puffs of vapor filling the air with mineral smells. The thriving and pretty town found here grew about the hotels and cottages first erected to accommodate visitors to these springs. It owes its existence to the enterprise of one man, Samuel Brannan, who took the little valley a solitude and has peopled it with a prosperous community of farmers and traders. His expenditures for improvements here were very large, and include such objects as vineyards, wine and brandy vaults, mulberry plantations for silk-culture, etc. He was also a prominent actor in the railroad. The planting of ornamental trees and shrubs about the springs was thought a doubtful experiment, by reason of the alkali and heat the springs diffuse through the soil. But the plantings throve slowly, and Calistoga is growing under the shadow of its grand mountain, which the plain mimics by a small isolated cone (Mount Lincoln) that rises from its centre. Soda and sulphur are the principal mineral constituents of the thermal waters, whose heat rises to the boiling point. In the hills near by are the remains of a petrified forest, the stony trunks of oak being quite numerous. When growing they were evidently buried by an earthquake shock, exposed to a watery solution of volcanic matter, which silicified them, and subsequently elevated again, and partly uncovered by the washing away of the enveloping earth. Mount St. Helena was once the centre of volcanic disturbance in this region, and threw its ashes and lava over a good part of the surrounding country. The hot springs at many points in the valley and hills, the pumice and obsidian scattered widely over the surface, the masses of volcanic rock observable, all indicate a time when this was a volcanic centre. And these indications extend northward at least as far as Clear Lake, some forty miles distant, where deposits of sulphur and a lake richly charged with borax are found. The earthquakes still felt occasionally through this region are not alarmingly severe. In December, 1859, a tremendous explosion was heard at Mount St. Helena, which shook the earth; but this the state geologist, Prof. J. D. Whitney, thinks may have been caused by heavy masses of rock in some of the subterranean cavities known to exist in these volcanic regions. During the winter a new hot spring burst out of the eastern side of Mount Lincoln, scarcely more than fifty feet above the valley-level, and has continued to puff away ever since. This circumstance excited less comment in the vicinity than the increased number of trout in the mountain-streams and the abundance of wild pigeons. Your true Californian is never much surprised or dismayed at anything. When the terrible earthquake at Inyo, in the southeastern corner of the State, was at its height, the survivors of the first shock amused themselves by inventing names for the various phenomena, the heaviest of the artillery-like discharges from the vicinity of Mount Whitney being called "the hundred-pound Parrott of the Sierra," while as the ground began to heave and shake again, the bold fellows would cry out, "There she goes! Brace yourselves!"

Mount St. Helena was ascended in 1841 by a Russian naturalist, Wasnossensky, who named it in honor of his empress, and left on the summit a copper plate, inscribed with the names of himself and his companion. This plate is now preserved in the museum of the California Geological Survey. The Russians did a good deal of exploring in California in early days, not alone for scientific purposes, but with some eye to commercial and political aggrandizement. They left their name at several points in the northern interior, including the Russian River and the lovely valley it waters, which opens north of Sonoma Valley and lies across the ridge to the northwest of Napa Valley. The tourist who is acquainted with these facts regards the country on the route to the Geysers with more interest.

Early in the morning a stage leaves Calistoga for the Geysers, distant twenty-eight miles. This "stage" is simply a very strong open spring wagon, seating nine to twelve persons. Last year it was not uncommon for half a dozen such wagons to make the trip daily. The road soon quits the valley, ascends a range of wooded hills to the northward, crosses it at a height of three or four hundred feet above the valley, and seven hundred and fifty above the sea, and descends to the northwest into Knight's Valley which is drained into Russian River. There are numerous creeks in this region, leading to many picturesque side valleys heading in the hills. Broad natural meadows are dotted with groves of oak, and in the spring months the green levels and slopes are spangled with flowers, including the blue lupin, larkspur, purple primrose, yellow poppy, and a profusion of buttercups and daisies. The streams run tinkling over gravelly beds, larks and linnets sing joyously, flocks of blackbirds chatter musically as they whirl in gusty flights together, and the clear air exhilarates like champagne. Mount St. Helena is kept to the right, revealing its sculpture boldly as it is neared, but never losing its magic tints. The ridges dividing a series of intervales are thickly wooded with oak and pine, with here and there a redwood astray, a madroña or manzanita, whose brown or red bark and waxen leaves make them very striking objects. If it is spring, big clumps of buckeye will thrust out their bristling spears of scented bloom. Where the soil is bare, it is red, except in the valleys, where it is black or brown, while the rocks are stained with lichens. Thus there is a constant feast of color, — gold and purple predominating in summer, emerald and red and violet in the spring, but always an undertone of pearly gray, which St. Helena's cone seems to give out as the key for the whole beautiful composition.

As the Geyser mountains are neared, the valleys narrow to ribbons, run into hills, and end in a dense forest glade, where lighter wagons are taken for the ascent. From this point teams are not allowed to travel in opposite directions; the road is too narrow and dangerous to pass. Hence the teams going out and in meet in this glade, composed of lofty firs in great part, and having the hushed air and soft carpet of a true forest. The summit of the first range of hills is about one thousand seven hundred feet above the station at its foot, or nearly two thousand three hundred feet above the sea, and the ascent is made in a distance of about four miles. These hills form the lower slope of Geyser Peak, which is three thousand four hundred and seventy-one feet high, and forms one of the triangulating stations of the United States Coast Survey, being plainly visible from the ocean and San Francisco. It is a conical peak, like all the dominating points of this range, and commands a magnificent view. The stage-road ascends its flanks very deviously, passing alternately through dense thickets of underbrush or bits of coniferous woods; then across deep gulches, watered with clear trout streams; then emerging into open spaces, and winding along the edge of a precipitous descent, opening far vistas of colossal scenery, rank on rank of diminishing hills thrusting up sharp tops of fir or pine, until these are lost in the blue gulf nearly two thousand feet below. Everywhere, except in the forest belts and thickets of brush, the more or less rounding hills of the Coast Range bear a luxuriant growth of wild oats. The clump or masses of tree verdure relieved against these golden slopes present an indescribably brilliant effect, which is enhanced by the dark blue of the chasms below and the purple or violet of the remote ranges beyond.

Resting the sweating horses for a few minutes on one of these wild harvest spaces, and looking about, the passengers have a view never to be forgotten. Across a gulf to the east rises the commanding bulk of Mount St. Helena. To the west and south descend the hills we have been climbing, and others beyond them, leading the eye to Russian River Valley, where the stream makes a sharp turn and can be traced on its gleaming course for many miles. The receding hills, with their shaggy coating of forest and chemisal, are softened with a violet haze. The valley shimmers in its heat, and through a cleft in the far blue wall of the outer Coast Range the sunny Pacific is seen melting into heaven. The air is wonderfully clear and luminous, lending the charm of its tints to the magnificent landscape, without obscuring it, so that we seem to be looking at it, almost dizzily, through a transparent medium which only reflects an image. Such a sight intoxicates the senses almost to pain. The world never appeared so lovely, nor our own nature so capacious and receptive. It is with a sigh of regret that we feel the wagon start and dash onward; but the extreme beauty of the woods is another delight. The madroña has become a tree, and its smooth mahogany limbs and large waxen leaves are rich beyond any other tree in the forest. Then the laurel and the bay, with their perennial green, the maple and the alder in moist places, and the blooming buckeye, fill up the spaces between the leather-colored columns of redwood and cedar, and the straight shafts of pine and fir, towering above all. As the road winds higher towards Geyser Peak, it leaves the forest and passes through a dense thicket of chemisal shrubbery, oak, laurels, small bays, and ceanothus. The last, called California lilac, is covered till late in the spring with powdery blossoms that give forth honeyed odors. Masses of stained and blackened rocks, serpentine, sandstone, and trap, rise here and there, giving the nearing summit a desolate look, which is increased by the few contorted pines that suck a feeble life from the crevices where they grow. A narrow ridge, called the Hog's Back, — just wide enough for the wagon, — connects two spurs of the range at this point, separating Sulphur and Pluton creeks. It is the parapet of a wall whose sides slope at a sharp angle a thousand feet, and riding over it at a high speed one looks into a chasm on either hand, catches breath, and hopes the harness and wheels may be strong. The Hog's Back, however, forms part of the old road which is not traveled now, except by daring tourists who insist upon going back by that route especially to enjoy a sensation. The new road keeps more to the flank of the ridge, and curves about precipices instead of crossing them. Both roads approach within two or three hundred feet of the summit of Geyser Peak, and then plunge suddenly down its farther and steeper flank to the cañon of Pluton River, on whose right bank are the Geysers. The greatest elevation either road attains is about three thousand two hundred feet. As the Geyser Hotel is one thousand six hundred and ninety-two feet above the sea, the descent is about one thousand five hundred feet. This is made on the old road in a distance of two miles. Foss, the proprietor of the road and stage line, and one of the celebrated "whips" of California, used to call this steep descent "the drop," and as he began it, would tell the passengers to look at their watches and hold on to their seats and hats. He would then crack his whip, and the horses — sometimes six to a wagon — would start at a keen run and make the distance in nine and a half minutes. There are thirty-five sharp turns in "the drop," and the road, just wide enough for the team, frequently hugs the edge of steep, rocky precipices, whose sides and bottoms made a concavity of bristling fir-tops, hiding the stream whose murmur comes faintly up. The new road makes the descent to the cañon of Pluton Creek, or river, by a longer route, with more curves, in a lighter grade; but is equally narrow, and follows closely, for long distances, the steep precipices that line the creek. Over this, too the teams are driven at a rate of speed frightful to timid persons unaccustomed to mountain stage-travel in California. But dangerous as these roads seem, not a single accident has occurred on them, for the wagons are kept very strong, the horses are of the best roadster stock, and the drivers masters of their trade. The great speed made, instead of increasing the danger, lessens it. Yet there are persons in almost every wagonful of passengers who pale and shrink as the vehicle dashes wildly down, and as they see below them, under the very wheels, as it were, the yawning chasms that seem to threaten death. Women sometimes sink into the bottom of the wagon, and hide from their eyes the spectacle so dreadful to them, that is so sublime to cooler heads and calmer nerves. When the wagon reaches the hotel, however, all its tenants have a half wild look, as if they had just come down in a balloon and were thankful it had "lit." Nothing can be more wildly romantic than the scenery of the Pluton cañon. On one side rises a steep mountain, rock-ribbed and clad with stately firs, mixed with evergreen oaks, bay trees, and madroñas; on the other side sinks a precipice into a deep gorge, crowded with a richer variety of foliage, through which are caught glimpses of a stream making foamy leaps over rocky rapids, or expanding into still pools, in whose depths fishes can be seen like images fixed in glass. Here a small brook comes tumbling down the mountain, creaming a mass of black rock a hundred feet high, which is margined with ferns, splotched with lichens, and shadowed by arching trees, out of which the cascade seems to leap. There, on the right, far across the cañon, other mountains rise, sparsely timbered with oak, yellow or green with wild oats, scarred with deep red gulches from summit to base, and — yes, actually smoking like a volcano from many an ashen heap or hollow. The air is charged with sulphurous smells, and as the sweating horses swing rapidly around the last curve of the road, by the last dizzy brink, we realize that there are the Geysers.

The Geyser Hotel is a lightly constructed framehouse, L-shaped, with double piazzas on all sides. It stands amid a grove of tall firs and massive evergreen oaks, on a narrow bench about one hundred feet above Pluton Creek, the mountains rising straight behind it. This creek is a tributary of Russian River. It heads up towards Mount St. Helena, and until it comes within the influence of the Geysers is a charming trout stream. Its banks and bed are extremely rocky. Huge boulders of granite and sandstone choke its course, and black volcanic masses rise in frowning cliffs by its side, sometimes softened with a drapery of vines, and bearing trees on their creviced tops. Great blocks of conglomerate, apparently formed *in situ* by the mineral constituents of the waters percolating through the diluvium, are also seen obstructing the creek. Occasionally it has cut through a bed of this conglomerate, which forms its banks. For all this ruggedness the creek is very picturesque, and has many spots of gentle beauty where the sun beams athwart quiet pools, and maples and pepper trees mix their gentle grace with the sombre foliage of fir and bay and evergreen oak. Pleasant paths wind along its banks under archways of green, where ferns and flowers thrive and coax the hand to pluck. Between the rocks round plats of tuft-grass make soft stepping-places. The quail is heard calling his mate in the thicket, and the robin chants his song at morn and eve in the tree-tops.

The best time to visit the Geysers is early in the morning, before the sun has risen above the mountain tops and drank up the vapors. From the red riven side facing the hotel, columns and clouds of steam may then be seen rising to a height of two hundred feet or more, obscuring the landscape like a fog just rolling in from the sea. The same phenomenon is visible, but in a less degree, towards night. It is pleasanter to take a good rest at night, to enjoy the concert of the birds in the grove about the house, listen to the soughing of the firs, the soft roar of the creek, and the distant puffings and gurgitations of the Geysers; and then from your bedroom opening upon a piazza, gaze out, as you lie with open door and window, in that balmy climate, at the keen stars beaming with their eternal quiet over that strange scene. Up before the sun, don an old suit, swallow a cup of coffee, and join the laughing party of tourists gathered about the guide on the fenced space before the house. Every one takes a "geyser pony,"— that is, a stout stick to help him or her over the rocks and springs,— and then all start down the trail, Indian file, to Pluton Creek. Before reaching it, the guide, who perhaps is the jolly landlord himself, points out a chalybeate spring of fine tonic properties, whose waters his guests imbibe, mixed with soda-water. The banks are charged with iron salts for a great distance up and down, and their solutions have given the earth its red tinge, and hardened the gravel-beds into a semi-metallic mass.

In curious contrast, at the crossing to Geyser Cañon, is the whey-like tint of the water in the creek, which for a quarter of a mile or more is affected by the sulphur discharges, some of which bubble up through the very bed of the creek itself. Thermal springs of various sorts are numerous along the creek, especially on its right bank, for several miles; but the most remarkable are those facing the hotel. The prevailing rocks are metamorphic sandstone, silicious slates, and serpentine. Their stratification is boldly exposed, and dips at a sharp angle to the line of the creek. Through the lines of fracture or cleavage, from the water's edge to a height of fifty or a hundred feet up the slope opposite, where the creek is crossed by a rustic bridge, numerous springs and steam jets escape, coloring the face of the slacking rocks vividly with the yellow, red, and white salts of sulphur, iron, lime, and magnesia that they deposit. The springs are of various temperatures, some of them exceeding 200°. One forms quite a large stream, and is led by troughs into a row of small shanties, where its steam is used for bathing, the bather jumping immediately after into a rocky basin of the creek two or three feet off, the waters of which are almost shockingly cool. Where no heated waters flow from the rock, the steam issues under a high pressure, intensely hot, and shrieking or hissing. From one hole, a foot or two wide, at the base of the bank, it escapes with a noise like that of a high-pressure steamboat "blowing off;" and this vent is appropriately called the Steamboat Geyser. For a hundred rods here the rocks are hot under the feet, and as they are also slippery with moist mineral salts, and puffing from numerous small vents, the spectacle they present is in sharp contrast to the sylvan beauty of the creek. Yet grasses grow in these heated rocks, out of the very salts, and one or two thermal plants dare to blossom at the edge and in the very breath of the hottest springs, whose waters are sometimes greened with low forms of microscopic plant-life, which also slime the rock where they overflow.

Following down the right bank of the Pluton for

a short distance, the trail turns to the right and enters a gorge densly embowered by shrubs at its mouth, but soon opening into the desolate regions of the Devil's Cañon. The nomenclature, like the scenery from this point, is all infernal, suggestive of Dante and his awful journey, except that the tourist hither seems to have reversed the course that Dante took, approaching Pluto's sphere from the region of Elysian beauty, instead of passing through that to these. Much of the nomenclature fastened to various points in the cañon is arbitrary and impertinent enough, and one wishes it were possible to see the place dissociated from all names that suggest superstition and cruelty. Climbing up a ledge that crosses the cañon, we suddenly gain a view of the principal Geysers. The gorge for half a mile up the mountain lies before us, a steep ascent, filled with steam and noise, its bare sides painted many colors, its bed obstructed with boulders, around and under which turbid waters gurgle and smoke; at the very head of all the apparent combustion and explosion an abrupt and tall cliff of red rock, bearing a flag-staff. The ascent of this gorge is toilsome but exciting.

Before the crusts of salt and sulphur and decomposed rock had been disturbed, and a trail marked out where the footing was known to be solid, the ascent may have been dangerous. It is certainly not so now, although to many persons very unpleasant. The hot ground under the feet; the subterranean rumblings; the throbs and thuds near some of the largest and most energetic steam vents; the warm, moist atmosphere, filled with ascidulous vapors, often charged with sulphuretted hydrogen; the screaming, roaring, hissing, gurgling, and bubbling of the various springs, — all contribute to make the scene as repellant to some as it is grand and exciting to others. Where the vapors are thickest, and the noises loudest, the guide says, "This is the Devil's Laboratory;" and so his Satanic Majesty gets the credit all the way for some of the most curious and instructive of the inner workings of that kindly power whose most terrible forces are instruments of good — manifestations of laws that operate through all time and space with impartial grandeur, without vindictiveness or hate.

There are no spouting fountains in the cañon, but numerous bubbling springs, that sink and rise with spasmodic action. These number a hundred or two, and are of varying temperature and constituents. A few are quite cold, closely adjoining hot springs, while others have a temperature of 100° to 207°. Some appear to be composed of alum and iron, others of sulphur and magnesia, while a few are strongly acidulous. Here the water is pale yellow, like that of ordinary white-sulphur springs; there it is black as ink. The mingling of these different currents, with the aid of frequent steam injections, intensifies the chemical action, the sputter and fuming, that are incessantly going on. These phenomena are not confined to the narrow bed of the gorge, but extend for a hundred or two feet up its sides, which slope at a pretty steep angle. These slopes are soft masses of rock decomposed or slackened by chemical action,

and colored brilliantly with crystallized sulphur, and sulphates of iron, alum, lime, and magnesia, deposited from the springs and jets of steam, which are highly charged with them. As the rocks decompose and leach under the chemical action to which they are subjected, the soft silicious mass remaining, of a putty-like consistence, mixes with these salts. Some of the heaps thus formed assume conical shapes. They have an apparently firm crust, but are really treacherous stepping-places. One of the most remarkable steam vents in the cañon is in the top of such a pile, fifty feet up the steep slope. It blows like the escape-pipe of a large engine. The beautiful masses of crystallized sulphur which form about it, as about the innumerable small fumaroles that occur along both banks, tempt one to dare to climb, and face the hot steam. The mass shakes beneath the tread, and is probably soft to a great depth. Wherever in these soft heaps a stick is thrust in, the escaping warm air soon deposits various salts. Of course a walk over such material is ruinous to boot and shoe leather, while the splash of acid waters often injures the clothing. Everybody stops to gather specimens of the various salts and rocks. The guide presents to be tasted pure Epsom-salts (sulphate of magnesia), and salts of iron and alum, of soda and ammonia. Few care to taste the waters, however, which rival in their chemical and sanitary qualities all the springs of all the German spas together. Perhaps the most remarkable of the Geyser springs is that called, happily enough, the Witches' Caldron. This is a black, cavernous opening in the solid rock, about seven feet across, and of unknown depth, filled with a thick inky liquid, boiling hot, that tumbles and roars under the pressure of escaping steam, emitting a smell like that of bilge-water, and seems to proceed from some Plutonic reservoir. One irresistibly thinks of the hell-broth in "Macbeth," so "thick and slab," and repeats the words of the weird sisters: —

"Double, double toil and trouble
Fire burn and caldron bubble."

A clever photographer, Mr. Muybridge, conceived

the idea of grouping three lady visitors about this caldron, with hands linked, and alpenstocks held like magic wands, in which position he photographed them amid the vaporous scene with telling effect. Another notable spot is the Devil's Gristmill, where a large column of steam escapes from a hole in the rock with so much force that stones and sticks laid at the aperture are blown away like bits of paper. The internal noises at this vent truly resemble the working of a grist-mill. Milton's hero is sponsor for another spring called the Devil's Inkstand, notable for its black water, specimens of which are taken off in small vials, and used at the hotel to inscribe the names of guests on the register. Dr. James Blake, who has read before the California Academy of Sciences several papers giving the results of his observations on the Geysers, says that the water of the Devil's Inkstand contains nine per cent. of solid matter in the form of soluble salts and sediment, the former being in the proportion of 2.7 per cent., the remaining ingredients being in the form of a dark black sediment. The matter has a thoroughly acid reaction, which it owes to the presence of free sulphuric acid. It would seem that a large portion of the soluble matter is composed of ammoniacal salts, probably the sulphate of ammonia. This salt, which rarely occurs in the natural state, has been found by Mr. Durand, another academician, precipitated in large quantities from the vaporous exhalations at the Geysers. Dr. Blake's analysis of the water of the Devil's Inkstand shows that about fifty per cent. of the saline ingredients consists of volatile salts, the remainder being salts of magnesia, lime, alumina, and iron. The presence of so large a quantity of ammoniacal salts in the water of a mineral spring is quite exceptional. These salts have long been recognized as occurring in the fumaroles, in the neighborhood of volcanoes, and their origin, particularly in such large quantities as at these Geysers, opens up some very interesting questions as to the nature of the strata from which so much nitrogenous matter can be derived. The sediment in the above water, in the proportion of more than an ounce to a quart, is probably some compound of iron and sulphur. Professor Whitney of the Geological Survey accounts for the black color and villainous smell of the water in the Witches' Caldron as follows: the iron held in solution comes in contact with water holding sulphuretted hydrogen, when an ink-black precipitate of iron takes place.

Wherever one treads, going up the Devil's Cañon, the step slips or crunches on some of the chemical products of these springs. It is a relief after a while to emerge from the heated vapors and sulphurous smells, and standing on the flag-staff cliff (called the Devil's Pulpit, of course), look down on the cañon and across to the hotel. Phenomena of the same sort, on a smaller scale, however, are visible on the higher slopes, and in the lesser gulches, up and down the creek. One place, called the Crater, a circular cavity of considerable depth, with a level, hollow-sounding floor, is evidently the site of exhausted thermal action, where the mineral constituents in the rock had all been slacked out, and the ground had sunk in; though about the lips of this "crater" one or two vigorous steam vents are still in operation, and sulphur continues to be deposited in fine needle-crystals. Half a mile below Geyser Cañon are a large sulphur heap, incrustations, and evidences of former activity, some heat still remaining in places. A ravine near by contains a clear hot spring, which was formerly built over with stones and sticks by the Indians, and the steam used as a sanitary agent. It is still known as the Indian Spring. Just without the rude wall inclosing it, runs a cold spring of excellent drinking water. Four miles up the Pluton Creek occur what are called the "Little Geysers," similar in character to the larger ones, except that they issue from a gently sloping hill-side instead of a deep gorge. The rocks and the chemical action are the same.

As to the origin of the phenomena we have been describing, it may be said that there are two theories, volcanic and chemical. Professor Whitney says (in his " Report of Progress," vol. i. p. 95) that there will be no difficulty in understanding them when we consider that they are displayed along a line of former volcanic activity, and where even now the igneous forces are not entirely dormant. " The dependence of the Geysers for their activity, in part, on the recurrence of the rainy season, indicates clearly that the water, percolating down through the fissures in the rocks, meets with a mass of subterranean lava not yet entirely cooled off, and, becoming intensely heated, under pressure finds its way to the surface along a line of fissure connecting with the bottom of Geyser Cañon; in this heated condition it has a powerful action on the rocks and the metallic sulphurets which they contain, especially on the sulphuret of iron everywhere so abundantly diffused through the formation, and so dissolves them and brings them up to the surface, to be again partly redeposited as the solution is cooled down by contact with the air." Professor Whitney adds that phenomena of the same kind as those observed at the Geysers, and sometimes even on a larger scale, are exhibited all through the now almost extinct volcanic regions of California and Nevada. Even on Mount Shasta the last expiring efforts of this once mighty volcano may be traced in the solfatara action still going on near the summit, and which is undoubtedly due to the melting snow finding its way down to the heated lava, or other volcanic materials below, in the interior of what was once the crater, from and around which a mass of erupted matter has been poured forth and piled up to the height of several thousand feet. We know, on other authority, that earthquakes have frequently been experienced at the Geysers, accompanied by loud noise. Two smart shocks on the night of February 20, 1863, were followed by the bursting forth of new openings of steam and boiling waters. Such an outburst, on one occasion, caused a gush of steam up the left side of the cañon so hot as to kill all the trees and shrubs in its course.

The chemical theory asserts that all the phenomena are ascribable to the action of water percolating through mineral deposits, and creating heat, expansion, and explosion by simple chemical decomposition, without the aid of a heated volcanic mass. The two theories may be harmonized, for the mineral matter is probably of volcanic origin, and whether it is heated before the water acts upon it is not very material.

In spite of the hot water, the steam and the saline deposits, vegetation flourishes far down the slopes of Geyser Cañon, about the margins, and in some of the very waters. The evergreen oak thrives almost within reach of the exhalations, and maples and alders are found on the banks of the creek close to some of the steam vents. A grass called *Panicum thermale* grows near the hot springs. Animal life dares to invade the scene, for dragon-flies of great beauty may often be observed, while birds build their nests and sing in the adjacent trees. Dr. Blake found two forms of plant-life in a spring having a temperature as high as 198°. These were delicate microscopic confervæ. In a spring having a temperature of 174°, many oscillariæ were found, which by the interlacement of their delicate fibres formed a semi-gelatinous mass. In a spring of a temperature of 134°, layers of filamentous green and red algæ were freely formed as the water flowed over the rocks. Unusual masses of oscillariæ flourish in the waters of Pluton Creek. Their presence in the highly mineralized waters of a spring with a temperature of 174° shows how great is the range of the conditions in which these forms of plant-life can be developed.

One returns to the hotel after a morning tramp through Geyser Cañon and along Pluton Creek with an enormous appetite, and is glad to rest for a few hours. Afterward, there are delightful strolls up and down the creek, and good trout-fishing for those who will go far enough. Deer and grizzly bear are to be had for the hunting in the mountains, — the grizzly sometimes without hunting. But the sportsman had better be accompanied by some one familiar with the country, unless he is a good forester, and can find his way without a path. A San Francisco lawyer was lost for several days on a hunting trip, and nearly starved to death before he was found. It was a roving hunter, of the true Leatherstocking sort, named Elliot, who first, of white men, found the Geysers in 1847. Coming suddenly to the edge of the cañon, he was amazed at what he beheld, and on returning to his companions told them, in his rough way, he had found the mouth of the infernal regions. Elliot fell in a fight with a tribe of Nevada Indians, not many years ago, a true border-hero to the last. The mountain over which he probably approached the Geysers, called Cobb's Peak, commands one of the grandest views obtainable in California. Northward, only fifteen miles off, lies Clear Lake, divided in two parts by the purple bulk of Uncle Sam Mountain, and surrounded by the rugged spurs of the Coast Range. On a clear day, one can see in that direction two hundred miles in an air-line, where the snowy crown of Mount Shasta, fourteen thousand four hundred and forty feet above the sea, floats in the sky like a fixed cloud.

Mount St. Helena and Napa Valley lie nearer at hand, and to the westward the eye takes in the Pacific Ocean for a hundred miles along the coast. Cobb's Peak can be ascended on horseback. The timber is not thick on the way, and many charming outlooks are obtained. Another scenical treat may be had by returning to San Francisco by way of the old road across the Hog's Back, to Ray's Station, and thence into Russian River and Sonoma Valley. Reaching San Francisco by this route the tourist will have gained a very good idea of the northern coast valleys of California, and the noble bay into which they partly drain.

GOLDEN GATE PARK.

Beyond the town, the bushy mounds between,
    Roll drifts of yellow wrinkled sand —
Uncrested waves, that dash against the green
    Like ocean billows 'gainst the strand;

But when the spring is soft, and winds are low,
    The shifting masses lie as still
As frozen banks of crusted moonlit snow
    That hide the hollow in the hill.

One way a mountain lifts its verdant crest
    Along a blue and cloudless sky;
On sloping pastures cattle feed or rest,
    And swallows twitter as they fly.

Below, around, the lusty lupin blooms
    In purple color, honey sweet;
The poppy's deep and golden cup illumes
    Each plat of grass or chance-sown wheat.

On rounded hillocks lustrous leafage shoots
    From laurel and from thorny oak,
And sprawling vinelets clutch with thirsty roots
    The soil no rain can ever soak.

A deep-set lakelet, greenly ringed about,
    Gems with its blue an open space,
Where yellow buttercups their beauty flout,
    And insects flutter o'er its face.

Through scenes like this the red and winding way
    Gives glimpses of the gusty town,
Throned on its many hills along the bay,
    Where far Diablo looketh down.

But westward, over sand-dunes ribbed and hoar,
    That deepen heaven's azure hue,
Are lines of snowy surf that faintly roar,
    Edging a sea that melts in blue —

A summer-shining sea, that slides and slips
    In silent currents through the Gate,
Where glinting sails of slowly moving ships
    For pilot or for breezes wait.

Northward, beyond a ridge of yellow sand,
    That hides the narrow harbor-way,
Rise headlands brown and bluff, whose summits grand
    Are islanded in vapors gray.

Below a line of arrow-headed firs,
    That stretches 'neath a strip of cloud,
The slope is softly greened, and nothing stirs
    But shadow of the misty shroud.

Peace broods where winds are fiercely wont to rave,
    To drive the sand like sleet before;
No sound disturbs the vernal stillness, save
    The surf upon the distant shore —

The faintly sighing surf, or linnet's song,
    Or music of the friendly voice,
Which gives to nature as we go along
    A charm that makes the day more choice.

CITY SCENERY.

The traveler who approaches San Francisco for the first time from the sea will not be charmed by its appearance, unless he has been fitted by a voyage of many months, like those early ones around Cape Horn, to welcome the sight of any land or town as beautiful. There is some beauty of form in the deeply eroded sandstone hills along the ocean where the surf dashes and roars constantly, and some richness in their tints of brown rock and yellow stubble under a summer sun and clear sky. There, as the ship enters a narrow strait leading to the bay, bold rocky cliffs on one side, a tall mountain on the other, the water covered with wild fowl, and the bay shores and islands coming into view ahead, the scene is picturesque and animated enough. But after leaving the rugged headlands that form the Golden Gate, and rounding a bold, russet-colored promontory, the gaze does not rest so much on these things as on the treeless sandy ridge, the formed lines of street cuttings which go straight over or through varying elevations, the mean architecture, the cold, monotonous gray of land and houses, marking the northwesterly extremity of the city. The long, gaunt peninsula, ribbed with outcropping strata of serpentine or sandstone, with long wave-like sand-dunes and rows of square wooden houses, remind one strangely of some monster skeleton of an early geological epoch, fossilized, and partly uncovered to the cold sea winds. It is only as another turn reveals the east front of the city, crowded with the shipping of all the world, covering more hills than Rome can boast, and flanked in the distance by greater elevations, that the metropolis of the Pacific presents a really attractive aspect. Situated on the extremity of a narrow peninsula, which divides the ocean from the bay, and built mainly on the inner slopes of ridges rising one above another from the water's edge to a height of from two hundred and fifty to four hundred feet, San Francisco has a bold and striking appearance. The silvery vapors lying like clouds in the distant intervales or mountain sags, the blended smokes of the city transformed by the sun into a softening haze, increases a grandeur of effect which is primarily due to elevation. While the local colors are brown and sere, except when the season of rain modifies them with verdure, vapor and smoke enforce a general tint of pearly gray, shading off into lilac on the higher and farther mountains, and harmonizing with the color of the bay, which only when calm reflects the pure blue of heaven. Though mist and smoke are mentioned, and either alone or together are seldom quite lacking, the upper atmosphere is usually sunny, giving a sparkle to the dancing water and a charm to the land.

The hamlet of Yerba Buena,[1] from which has de-

[1] Named from a sweet smelling indigenous plant.

veloped in a quarter of a century the present city, occupied a gentle declivity between the hills and a crescent line of beach whose horns were bluff promontories. But the pretty cove known to old whalers and pioneer gold-hunters has been filled in to a line drawn straight from point to point, forming several hundred acres of level land, which is now thickly built over and constitutes the commercial heart of the city. Clark's Point and Telegraph Hill, the northerly promontory of the old cove, have been cut away until they present a sheer precipice of brown siliceous sandstone, nearly two hundred feet high, on the dizzy verge of which rows of houses stand in bold relief against the sky. On the farther side this declivity slopes down to sand-hills and dunes that stretch along the bay-entrance for several miles and lapse at last into the sea-beach on the western side of the peninsula. Rincon Hill, the southerly point of the cove, was a less elevated bluff, covered with beautiful shrub oaks, laurel, and ceanothus; but this has been built over, partly cut down facing the lower bay, and quite cut through by a leading street which makes an excavation seventy-five feet deep with a talus of garden mould, trees and plants, the *débris* of ruined homesteads, and a crest of dilapidated houses toppling to their fall in a desolate way. The hills west of the cove, where they have not been quite leveled, filling up ravines and hollows, have been cut through by an arbitrarily rectangular street system, which may be taken as a good type of the invincible but tasteless energy of the pioneer builders, who would rather go rudely over a difficulty than gracefully around it. The resultant inconveniences of steep ascents for man and beast, of dwellings left perched high in air, of repeated expense to modify early blunders are partly compensated by the fact that many of the streets have the most picturesque vistas. Looking various ways one sees in the perspective villa-crowned cliffs, the craggy peaks or rounded contour of the peninsular hills, the straight blue ridge of San Bruno to the southward, the Golden Gateway cloven through beetling precipices, the dromedary-backed islands of the bay, or the rumpled folds of the Alameda mountains rising beyond the east side of the bay to a height of from fifteen hundred to two thousand feet, treeless except for the luxuriant groves of oak and laurel hid in their deep ravines, and lorded over by the lofty peaks of the Mount Diablo range lying beyond. The streets leading east and west give the passenger visions of the morning or evening sun, rising or setting in glory over landscapes that seem almost a part of the city. In most great cities nature is shut out by the walls of brick and mortar; but in San Francisco she always asserts her presence, if not her influence. The chief money mart, California Street, abuts upon a lovely picture of water and mountain and sky, at one end; while at the other end, down the steep flank of a high hill, the setting sun shoots his golden arrows and trails a robe of crimson cloud, glorifying the street even to the common gaze amid all its common houses.

Climbing to the top of this delectable hill, and of Russian Hill near by, some three or four hundred feet above the tide, we take in the whole topography and scenery of this fortunate city. The peninsula, twenty-four miles in length, and at its northerly end only about four miles wide, is made up of high sandstone and serpentine hills, both ridged and tumular in form, alternating with sandy knolls or long stretches of shifting dunes, and sometimes separated by grassy valleys, shrubby ravines and elevated plateaus. On one side is the blue Pacific, breaking in foam upon a long sandy beach; on another the bay, laving the city front and following the many indentations of the inner shore-lines beyond. If we look northwestwardly, we see the steep bluffs and rocky headlands six or seven hundred feet high, a deep reddish brown in color, with green slopes above that terminate in the sharp but handsome peak of Tamalpais. This mountain, about two thousand six hundred feet high, is only a few miles from the city, and rising so abruptly from the bay level is a prominent landmark in every direction for long distances. It is the terminus of a peculiar straight ridge which ascends gradually from the ocean side like an inclined plain, forming one of the ranges of Marin. Thickets of chaparral give it a dark color, according to the amount of humidity or sunshine in the air and the strike of the sun's rays. Under these shifting influences its tints are infinite. In a perfectly clear atmosphere, its local color comes out strongly, and it seems not one fourth as far away as usual. A very slight haze clothes it in a tender violet and sets it farther off. If mists are rolling in from the sea, they circle about its top, and lie in its hollows like fleecy clouds. A person who stands on its summit at such a time sees below him nothing but a billowy ocean of silver vapor, and enjoys in safety the spectacle that aeronauts attain only by perilous flights. If the mists are absent, the gorges on the northerly side are seen to be filled with noble groves of the redwood, fir, laurel, madroña, and other trees characteristic of the Coast Range; and there will be, far down, intervales of yellow stubble, relieved by clumps of dark green live-oaks and blooming masses of buckeye and

MOUNT TAMALPAIS.

ceanothus. The eye also takes in a fine panorama of ocean and bay; of the gray hilly city, and its environments of richly colored mountains; of the valleys opening to the northward and the Sierra Nevada in the east. But the sunset aspects of Tamalpais, from the town, are its peculiar glory. These are so rich and yet so tender, like the verse of Tennyson, that they defy description. A very dying dolphin of mountains is Tamalpais. No wonder that it is the love of local poets and painters, and that enthusiasts like Stoddard and Keith have made it a very mount of inspiration.

Looking northward, from one high point within the city, we see the islands of the bay. Alcatraz, a great brown turtle with red brick forts upon its back, bristling with cannon from all its steep shore-batteries, which have displaced the beautiful pelicans that gave the island its name. Angel Island, whose cone-like top is nearly eight hundred feet above the tide, a giant mound of grass and flowers in the wet months, of brown stubble in the dry; Goat Island, a dark olive green with its chaparral crown; and smaller islands farther off, where the upper bay pushes its narrower channels between low, mound-like hills covered with wild oats, to meet the yellow discharge of the river that winds lazily through the broad prairies of the interior. In this direction the vista ends with the high ranges of Sonoma and Napa, and Mount St. Helena, sixty miles off, lifts its peak of slaty gray over four thousand feet. Looking eastwardly, across the bay, which is here about five miles wide, we see at the base of the Contra Costa range the Alameda Valley, well deserving its soft Spanish name, for its gentle slopes are partly covered with dense groves of the California live-oak (*Quercus agrifolia*), quite uniform in the rounded masses of their foliage and their stout gray trunks, though curiously varied in botanical character, often loaded with bunches of mistletoe, and planted with an orchard-like regularity, opening on vistas of water, meadow, and hill. Here the milder climate permits a luxuriance of native flora which is in marked contrast to the rather limited growths of the sandy and windy peninsula, where, within the city limits, the sheltered spots that were once verdant enough, have been mostly buried by the leveling process and covered with buildings. On this favored slope a couple of miles wide and ten or twelve long, half a dozen oak-embowered towns nearly join their suburbs and dot the lesser heights behind them with pretty villas. Chief of these are Oakland and Alameda, which are nearly conterminus for six or seven miles, ending northerly in the charming vicinity of Berkeley, where the State University is growing with noble promise amid groves of oak and bay and laurel, by the margin of a bubbling brook, — a scene destined to be as classical in letters as it is already lovely by nature. The Alameda shore commands a grand view of the bay, the city, the islands, the Golden Gate and its sentinel Tamalpais, and even of the ocean beyond. The Contra Costa or Alameda mountains rise abruptly above it to an average height of fifteen hundred feet, deeply eroded from summit to base, treeless, except for the beautiful groves smuggled in the winding gorges and passes, which are not visible from the city. At intervals of a few years, a light snowfall robes them for half a day in winter — a spectacle of wonder in this mild region where the word winter calls up no ideas but those of needed showers, of verdure, and of bloom. Behind the Alameda hills rises the double cone of Monte Diablo, very near to the view, but separated from the hills named by the San Ramon Valley, and distant from the city easterly about thirty miles. This peak is three thousand eight hundred and fifty-six feet high. Rising from the centre of a wide basin which runs into the great valley of the Sacramento and San Joaquin, and being the most elevated spot in this region, Monte Diablo looms up in the perspective of every view in all directions around it, and is one of the most familiar landmarks to the citizen of San Francisco, who sees it daily and almost hourly. Its dark blue mass lords it nobly over the brown hills of Alameda, and when it takes on its snowy cap for a few days in the rainy season it

MOUNT DIABLO

is more peculiarly prominent. It is a great sun-dial, for the stages of the coming or going day are marked in bands of shifting color upon its top. Around its base, fertile valleys swell to meet its foot-hills as if they would embrace it, and hold a score of thrifty towns. From its summit one of the most extensive and beautiful views in the Union can be obtained. The great plains of the Sacramento and San Joaquin, stretching from the northeast to southwest nearly three hundred and fifty miles; the rivers of the same names winding their yellow currents from north and south, meeting at the head of the upper bay; the vast bulk of the Sierra Nevada with its snowy crest, along the eastern sky, from Lassen's Peak at one extremity to Mount Whitney at another; the isolated "Buttes" of Marysville in the centre of the Sacramento Valley; the line of the Coast Range from Mount St. Helena on the north, to Mount Hamilton, four thousand four hundred feet high, at the south, broken into lesser spurs around the bay; the whole scenery of the bay itself, the city, the Golden

Gate, the ocean beyond, — all this magnificent panorama, in clear weather, lies spread out before the spectator on the summit of Diablo. The area included within the bounds of this view is probably not less, according to Professor Whitney, than forty thousand square miles; adding what can be seen of the ocean it is much more. It might well have been on such a commanding height as this that the enemy of mankind tempted the Saviour; and an early Spanish legend, to which the mountain owes its name, actually located here a terrifying appearance of the devil to a party of explorers. This legend would seem to indicate a belief that the mountain is of volcanic origin, as it has been said to be by some writers; but it is simply a grand mass of metamorphic sandstone, flanked by jasper, shales, and slates, with limited coal-beds at its base and deposits of cretaceous fossils. The gap between the two peaks is eight hundred feet deep, and the north peak is nearly three hundred feet lower than its companion. From certain points of view the two peaks are brought into line and have the effect of a single perfect cone. Seen from the upper bay or river, the mountain seems to rise in this shape directly from the water's edge, and is very imposing in its near bulk. The ascent of it from any quarter, with the ever expanding outlooks revealed, is full of picturesque charm. The nearer scenery of the foot-hills and lower flanks — embracing graceful wavelets of harvest-land, melting into level spaces, deep gorges filled with evergreen growths, sandstone cliffs weathered into fantastic forms, and bits of charming brooks and grassy springs — is itself a treat to the lover of nature. Sunrise and sunset are the best hours for visiting the summit. At the former, the air is clearest, and one gets the widest view, besides the glorious spectacle of the great round orb flashing up above the crest of the Sierra, bringing its highest peaks of snow into sharp relief. The shadow of the peak is thrown in a pyramidal form over the whole country to the west, across the Alameda hills, the bay and peninsula of San Francisco, and into the ocean beyond, forty miles in length, — a dark bluish triangle of shade that shortens slowly as the sun rises higher and higher, that withdraws by almost imperceptible degrees from the ocean, from the peninsula and bay, from the Alameda range and San Ramon Valley, up the flanks of Diablo himself, and there at last quite disappears. At evening this spectre of the peak is reversed, falling over the San Joaquin Valley, up the Sierra, and even into the sky, at first gradually lengthening as the sun sinks lower in the west, and then losing itself in the general twilight and darkness of his disappearance. Looking seaward then, we observe the myriad lights of the city, if no fog obscures them, and on the distant Farallone Islands the flashing of the beacon set to warn mariners.

Returning again to our hill in the city, one overlooks the undulations of the metropolis all around him, and has a vivid sense of the abounding energy, increased by the stimulus of a dry and equable climate, which created the place from nothing. Over the populous levels to the west and south, which lie like gulfs between California Street Hill and the Mission Hills, hang vapors and smokes that the evening sun transforms into beautifying haze, like those gauzy veils that women wear to enhance their charms. The Mission Hills bound a plain where stands Dolores and still rings its centuried bell, in the heart of the busy community which has succeeded its primitive congregation of simple savages. These hills, eight or ten hundred feet high, dividing two extremities of the city, are brown and barren enough near at hand, though always graceful with their cap-like peaks, and richly dight with buttercups and poppies in the spring, or with purple at all seasons when the setting sun makes them aflame. Farther in the same direction the high walls of the San Bruno Mountains are drawn in darker purple along the sky, the bristling fir trees scattered on their summits, distinctly visible, calling to the citizen's mind memories of the solemn, sonorous woods that look upon the sea. From the base of these mountains, which mark the breaking down of the Santa Cruz Range, stretches around the southerly end of the bay a margin of fat valleys, rich in grain and fruit, embracing those of San José, Santa Clara, and curving eastwardly again to Alameda. As the sun descends, the bay begins to reflect the tints of the sky. Shadows fall into the hollows of the city, and crawl up the slopes of the Alameda hills, beyond which the top of Diablo is ruddy with the last glow of day.

But before day closes let us descend to an intervale lying farther west, and thence climb the ridge which is crowned by the monumental peak of Lone Mountain, around whose slopes, looking both towards the city and the sea, all the worry and passion and pride of the hard metropolis sink at last into the grave. The noisy town on one side, and the still blue Pacific on the other, of these thousands who have gone before, are apt emblems of the lives they led, and the peace they have found. The city thins into scattered hamlets, that are lost in drifting sand; and beyond one sees the ocean, hears the faint roar of its surf, and, when the air is clear enough, catches glimpses of the Farallon Islands, thirty miles away, where the imagination pictures the sharp gray cliffs populous with seals, gulls, and murres. Among the sand, on every hand are hillocks of green shrubbery, with intervales of grass, hollows filled with ceanothus thickets and groves of stunted live-oak, and even a lakelet or two, where a great park is in progress of creation. The mists that often roll in over the seaward slope maintain an olive-tinted verdure through the long rainless summer; but the landscape, except on the sunniest days, when little or no wind blows, is sombre and melancholy. After the rains begin, in October or November, and thence until May or June, extensive thickets of lupin and ceanothus, encroaching on the drifting sands, take on a brighter green and burst into profuse bloom, blending their tints of lemon and purple and blue, and scenting the air with honeyed sweets for miles. Orange-colored poppies contest the open spaces with shining buttercups; the grassy slopes of the San Miguel Mountains are dotted with cattle; the far ocean is blue and sunny, creeping slowly upon the beach of white sand.

At this season, also, the more distant landscape southward, and on the eastern side of the bay, as well as north of the Golden Gate, takes on a light pea-green, to which the vaporous air rising from the water gives a soft gray tint as seen from the city. No color can be imagined more delicate through the day, or more lovely in its softening tint of violet at evening. And a constant phenomenon of sunset is the flush of pale pink far up the eastern sky. When night settles down, the view of the city from the hill-tops, illuminated by long processions of gas-lights paling the wan stars above, is singularly impressive. Looked at from the bay side, approaching on the water, the night aspect of the city is still more striking; for details are lost, and only the thick lights as they climb the hills are seen as so many ruddy stars against a dark background, — those on the wharves and shipping casting long, tremulous reflections in front.

How fortunate is San Francisco in these picturesque surroundings and effects! How fortunate again in the high points within her limits and suburbs, which command one panoramic view from ocean to mountain, around the shores of the peninsula and the bay. Scenically, there is no other American city so happy. And then the climate of summer and spring, whose means of temperature vary only ten or twelve degrees, the seasonable succession of dry and wet, of russet and green, the alternations of clear and misty air, are circumstances which give a peculiar variety to the scenic effects. The city landscapes have their moods, as though they were human. When the atmosphere is transparent and still, the town glows with a mild heat; the bay is like blue satin with shadings of pink; the mountains on every side are wonderfully bold and near, revealing every detail of their sculpture as well as the strength of their local color; the sand-dunes lie still against the bluest sky; and the ocean wears an expression exquisitely dreamy and gracious. Sparkle and motion, without loss of clearness, succeed the languor of such a day when a light wind blows, ruffling the bay, and giving a louder tone to the surf. Shining masses of vapor may roll inland, but they are cumulus, not sheeted, and rest peacefully in the bosom of the hills, where the sun makes of them a splendor. But when a true fog comes, it envelops everything in its gloom; chills the soul and body; gives a cold gray look to the city, and makes a dolorous way of the Golden Gate, where it pours in as if it were a troop of sad spirits. Such contrasts are full of poetic suggestions, and unite to make a character for our city scenery, changeful yet not capricious, full of charm and compensation.

THE FAWN ON 'CHANGE.

(CALIFORNIA STREET, SAN FRANCISCO.)

It stood amid an eager crowd
    Of brokers on "the street"—
A mild-eyed fawn, led by a thong
    That checked its impulse fleet.

Its pretty hairy sides were brown,
    Its ears were large and soft,
And lightly moved its little hoofs
    As though they trod a croft.

A cruel hunter killed its dam
    While browsing in a glade
Of redwood hills, and saved the fawn
    For profit in a trade.

And so it came to Mammon's court,
    Where fearlessly it stood
As though beside its dam again
    Within its native wood.

How many features hard and stern
    Relaxed before its grace!
How many hands were gently laid
    Upon its pretty face!

Like guileless babyhood it touched
    Those avaricious men,
Who stopped to meet its lovely eyes,
    And turned to look again.

The hidden springs of feeling, choked
    By sordidness so long,
Welled up within them as they gazed,
    And bubbled into song —

A quiet song, that filled the soul
    With memories of days
When eyes as soft, of girls as pure,
    Beamed on them love and praise;

With memories of days afield,
    When nature, for the boy,
Had still a charm that made him thrill
    With health-bestowing joy.

And as they pass along they see,
    Far down the avenue
Of busy trade, a purple line
    Of hills against the blue.

Where bay and oak the gorges fill,
    And velvet shadows lie,
And birds uprising from the wave
    In lazy circles fly.

They smell the wild rose in the street,
    And far beyond the town
They seem to wander, where the lark
    His melody pours down.

SANTA CRUZ MOUNTAINS.

EVERYWHERE in California the Spaniards or Spanish-speaking Mexicans left the soft nomenclature of the most musical language of Europe. They saw much here to remind them both of Spain and Mexico,— in the lofty Sierras capped with snow, in the broad valleys, in the rich contrasts of russet and green tints under a cloudless sky. Hence it was natural to transfer to the new land many of the names familiar in the old. The religious sentiment of the Mission Fathers and their followers led them to add names of sacred meaning, equally musical. From these two causes it results that California rejoices in a nomenclature which both recalls visions of Old Spain and revives the religious traditions of the Middle Ages.

The names of the mission establishments were extended to the adjacent country, to valley, mountain, or stream, as in the case of San Buenaventura, Santa Barbara, and many other charming localities of mellifluous title. In this way the name of Santa Cruz was extended from the Mission of the Holy Cross, founded in 1791, to the noble mountains that rise behind it, overlooking the ocean. The Santa Cruz Mountains are simply a conspicuous spur of the Coast Range, beginning a few miles south of San Francisco, and extending fifty or sixty miles parallel with the trend of the coast. On the east side of this spur lie the extensive valleys of San José and Santa Clara, which skirt the lower end of San Francisco Bay, and are bounded by the Alameda ridges of the Coast Range farther inland. Between the Santa Cruz Mountains and the Pacific there is only a narrow strip of terraced soil, marking the recession of the ocean at different periods as the land was elevated, and leaving fertile plateaus where anciently rolled wave and tide.

These mountains have a base about twenty-five miles wide, and an elevation of from two thousand to nearly four thousand feet, including several characteristic peaks. As they consist chiefly of sandstone, they have been eroded during tens of centuries, by the copious winter rains of this climate, into most picturesque forms. Their slopes are channeled with deep ravines, their crests cut through by numerous passes, dividing conical or tabular summits, and their bases, spread out in tumuli-like foot-hills, gradually sink into level benches or valleys. At one place near Santa Cruz the sandstone of the lower slopes has weathered into forms curiously resembling columnar ruins and castellated piles. Along the ocean it is cut into cliffs and walls that loom gray and resplendent, and against which the surf dashes and roars without rest; though the harvest may be yellow to the very edge above, where bloom purple flowers fed on the spray. But the glory of the mountains is their magnificent forest of redwood, which clothes all their upper flanks in perennial verdure, and grows lustily in the congenial sandstone soil and sea fogs, without which these trees do not flourish, if indeed they can exist. Pines, firs, and oaks there are also, but these we have seen elsewhere, while the redwood has its habitat only in portions of the Coast Range. Leaving the sandy and treeless peninsula of San Francisco, skirting the marshy shores of the lower bay, and crossing the fertile valley of San José, whose level surface of harvest, orchard, and vineyard stretches to bare, brown hills, which are relieved only in the southwest by blue mountains, it is a fine contrast to dash up the umbrageous flanks of the Santa Cruz, through clouds of red dust it may be, but also through such forests as one sees nowhere out of California.

There are two ways of going over the range from the inner valleys, and each has its special features. Turning to the westward from the pretty oak-nestling town of San Mateo, which lies in a narrow vale crowded between the bay and the Sierra Morena branch of the Santa Cruz Range, the stage leads up the San Mateo Creek, — a little trout-stream embowered with chestnut oaks, with densely-leaved and aromatic bay trees, with tall, straight alders rooted in the very water, and with many flowering shrubs, its lower banks curtained by hanging vines or edged with mosses and tufted grass. What an exquisite sheltering from the summer glare outside, which burns down on rolling hills yellow with grain or stubble, where only rare clumps of shrub oak or buckeye relieve the sight with patches of grateful shade, or the madroña shows its smooth, ruddy bark and lustrous waxen leaves, dwarfing the not dissimilar manzanita. Leaving the creek, and going steadily upward, the road curves among lofty rounded hills, that wave their green or yellow harvests in rippling lines along the sky edge on either side, producing a most curious effect of color and motion. Nothing can be softer than the myriad wavelets of light and shade, while the breeze-tossed grain rolls ceaselessly against the blue heaven far above the eye. And so we reach a narrow summit some two thousand feet above the valley and the bay, and have a rich prospect as we gaze down upon them and beyond to the treeless ridges of the Mount Diablo range. Immediately descending again, we enter a long and narrow pass, cut deep in the western slope, which leads almost straight to the Pacific, opening a free view of the blue ocean and its white crescent lines of surf beating upon the green beach of Half Moon Bay. This gorge is comparatively treeless; but its lower slopes are cultivated, and along the black, loamy banks of its little stream are patches of the yellow primrose and sweet-brier, of spotted tulips, golden poppies, and purple lupins. The moist sea air keeps the grass always green, and we seem to have suddenly reached another climate than that of the warm and dry interior, with its prevalent summer colors of russet and chrome. A most exhilarating dash, with the ocean always in view ahead, brings us to the shore, where we turn southward through uneven benches cultivated to vegetables and grain, hugging the rugged hills on one side, and gazing with unflagging zest at the continuous lines of surf on the other. The day's journey ends at Pescadero, a white, snug, New England looking village, on the level banks of a creek by the same name, which puts down from the redwoods to the sea and empties through a rolling pasture-land two miles from the town. Judging from its name, Pescadero must have been, in Mexican times, a favorite fishing resort; indeed it is yet, for the numerous streams in the vicinity abound in trout, other varieties of fish coming up with the tides from the sea, and the very surf on the shore containing a peculiar viviparous fish, the catching of which is amusing just in proportion to its uncertainty. The little oblong valley where the village stands was once a salt marsh, and is still a marsh at its seaward extremity. Shut in by long, wave-like hills, which are always green with chaparral thickets where they are not made into hay and grain fields, its proximity to the ocean is announced only by the morning mists and by the distant roar of the surf, which reminds one at night of the solemn monody of Niagara. The rolling upland that leads out to the Pacific is a rich pasture, and forms part of a famous dairy range. Among the grass, within reach of the drenching spray, wild strawberries are plentiful in the season. Suddenly this pasture edges upon a steep bluff that overlooks the ocean. At the foot of this bluff, and shallowing out to the rocky bar, lies a beach composed of wonderfully clean and beautiful pebbles, including jasper, agates, carnelians, and other siliceous stones, derived from an adjacent stratum of coarse friable sandstone, and worn lustrously smooth by constant rolling on the surf, which flings them back in huge windrows daily. The opaline, pearly, amethystine, amber, and ruby tints of these pebbles are enhanced as they lie wet at the edge of the surf. One seems to have fallen, like Sinbad, upon a Golconda of gems. The labor of picking out the most beautiful of the small pebbles, which may not be larger than a pea, is very fascinating. People go to the beach to stay for an hour or two, and remain all day, reluctant to leave at last. Sober men of business, with hard lines of care on their faces, who put a monetary value on time, give themselves up to the beguilement almost as willingly as women and children. Groups of both sexes and all ages can be seen lying prone for hours, scratching with their hands for rare stones, shouting with pleasure when successful, or with pretty alarm when the surf pushes hissingly up to where they lie, leaving behind iridescent bubbles and brighter gleaming pebbles. As the surf breaks and foams over the flat rocks running out from the beach, it has a singularly reverberant yet soothing sound, varied by the thunderous roar that comes at intervals from the cliffs near by, where the spray tosses up to a great height. In the pools among the rocks, at low tide, one sees numerous beautiful polyps, grafted on the rocks like living chrysanthemums of the sea — animal flowers indeed. Upon neighboring benches of sand the pearly shell of the abalone is found, with many handsome varieties of algæ. Only five miles from this beautiful beach begin the superb forests of redwood which stretch up the western flanks of the Santa Cruz Mountains. A drive of three miles into the hills beyond Pescadero, the narrow roadway frequently embowered with willows, cottonwood, alder, and bay trees, takes us into the very heart of a dense and solemn grove, whence sunlight and sound are alike excluded, and the slightest motion or chirrup of a tree-squirrel seems a disturbance. The redwood is a species of the same genus (*sequoia*) to which the Big Trees of Calaveras and Mariposa belong, and rivals the latter in magnitude as it resembles them in general appearance. Nowhere else than in California is any species of this genus found, except as a fossil relic of a past geological epoch. But the redwood surpasses the Big Tree in general effect, because, as Professor Brewer says, it frequently forms the entire forest, while the Big Tree is nowhere found except scattered among other trees, and never in clusters or groups isolated from other species. In the graphic words of Professor Whitney: "Let one imagine an entire forest, extending as far as the eye can reach, of trees of from eight to twelve feet in diameter, and from two hundred to three hundred feet high, thickly grouped, their trunks marvelously straight, not branching till they reach from one hundred to one hundred and fifty feet above the ground, and then forming a dense canopy, which shuts out the view of the sky, the contrast of the bright, cinnamon-colored trunks with the sombre deep yet brilliant green of the foliage, the utter silence of these forests, where often no sound can be heard except the low thunder of the breaking surf of the distant ocean, — let one picture to himself a scene like this, and he may, perhaps, receive a faint impression of the majestic grandeur of the redwood forests of California." Some of the redwoods in the forests near Pescadero measure from fifteen to twenty feet in diameter. Near Santa Cruz there is a grove containing equally large trees. Members of the Geological Survey have reported trees in the northern part of the Coast Range from twenty-five to thirty feet in diameter, and three hundred feet high. A hollow redwood stump exists near Eureka in which thirty-three pack-mules were stabled together. Near the summit of the Santa Cruz Range is a hollow tree in which an old hunter made his convenient home for a long time. Professor Whitney relates that during the strong winds of 1861-62 redwood logs drifted out to sea along the northern coast in such immense numbers as to be dangerous to ships one hundred and fifty miles from land. They were afterwards thrown ashore in great piles, and on being measured were found to vary from one hundred and twenty to two hundred feet in length; while one of two hundred feet was ten feet in diameter at the base, and another of two hundred and ten feet, was three feet in diameter at the little end. In a thick forest of such giants the soil is kept moist and cool, and supports a luxuriant undergrowth, including ferns and delicate flowers. Animal life prefers the warmer and brighter regions of the oak, on the slopes or in the valleys, and the thickets of the comparatively treeless hills; though at considerable elevation the redwoods are tenanted by the grizzly bear, which is sometimes more than a match for the luckless sportsman or traveler who encounters him.

The half day's drive from Pescadero to Santa Cruz, along the coast hills, and often over the very beach, is most exhilarating and picturesque. Always the rumpled folds of the bare sandstone mountains on one side, and the sunny surf and rolling ocean on the other, with occasional passages through ravines canopied by evergreen oaks and laurels, glimpses of white sails along the watery horizon, and precipitous outlooks over reefs where ships have been wrecked and their crews lost. A whale that is common off the coast, and is often pursued and captured by hardy men in small boats, who make this pursuit a business, is frequently seen blowing up his shining fountain of vaporous breath. At Pigeon Point there are odorous reminders that here leviathan is brought to the martyr's trial of fire for the good (oil) that is in him.

Santa Cruz stands on a triple terraced plain between the mountains of the same name and the lovely Bay of Monterey. Two long promontories jutting out about ten miles from the main-land, and about twenty miles apart, form the circular bay which is named from the old town on its lower side, Santa Cruz being on the upper side. The portion of sea thus inclosed is more delicately blue than the open ocean, and usually more tranquil. Nothing can be more graceful than the bent bow line of its glistening sand beach, whereon the surf breaks gently, accenting with its whiteness the tender blue of the water beyond. Behind the broad terraced valley which margins the bay, and which is dotted with groves of live-oak disposed in an order almost artificial, rise the mountains above a tier of foot-hills, to a height of three thousand feet, dark with their forests of redwood and fir, but taking on in certain conditions of the atmosphere deliciously soft tints of purple and violet and gray. It was in the midst of this noble landscape, equal to anything on the shores of the Mediterranean, that the Franciscan friars founded several of their earliest missions, raised the towers of their picturesque churches, which recall Castile and Granada, and labored to convert the simple aborig-

ines. The ascent of the mountains from this side affords a series of grand views, including deep gorges, outcroppings of gray rock, vistas of red-trunked conifers, vapor-girdled peaks, undulating valleys, winding streams, oak-embowered villages, and deep blue ocean. Above the crest of the range is the dark peak of Loma Prieta, luminous at sunset in a rich purple haze. At last the redwood forest completely shuts off the scenery on the western slope, and as we go eastward the next outlook is upon the oak-covered hills and golden valleys of San José and Santa Clara, bounded again by the bare brown mountains of Alameda, which skirt the inner shores of San Francisco Bay, and stretch southward to the connecting ridges of the Gavilan. If the season is spring, all this region will be clad in a garment of light green, having an undertone of that soft gray so loved by painters, and variegated by wild-flower patches of every color, while silvery clouds will move idly in mid-heaven, casting their shadows over the landscape. Such are the contrasts of a climate which has two seasons, of a land

LOMA PRIETA

Where birds ever sing, and summer and spring
  Divide the mild year between them;
Where the light-footed hours are told by gay flowers
  That need no hot-house to screen them;
Where there's gold on the plain, in the ripe waving grain,
  And gold in the far purple hills;
Where the tall, sombre pine giveth place to the vine,
  And the bee his sweet treasury fills.

AUTOCHTHONES.

No bronzed Apollos of the wood
   Those simple folk of El Dorado,
Who peopled once the solitude
   From Shasta to the Colorado.

But, short of stature, plain of mien,
   And lacking all the sculptured graces,
They still were part of every scene,
   And song and science seek their traces.

No monument of art arose
   Where once they dwelt in densest numbers;
The curious modern only knows
   By kitchen-heaps the tribe that slumbers.

Or, raking in the blackened soil,
   He finds the tips of spears and arrows,

Wrought by the ancient artists' toil
   To slay all game from man to sparrows.

Yet, artless as they were, and still
   As history will be about them,
They did their Mother Nature's will,
   And Nature could not do without them.

They were the Adams of the land,
   Who gave to hill and vale and river,
To every tranquil scene or grand,
   The titles that recall the giver.

While soft Solano spreads her plain,
   And lifts his head, tall Yallowballey —
The vanished people will retain
   A monument in hill and valley.

Yosemite their name inscribes
   On cataract and granite column;
And Tahoe murmurs of their tribes
   Among her peaks and forests solemn.

THE FIRST PEOPLE.

THESE sketches of scenery in California would be incomplete without some reference to the primitive people who once enjoyed that scenery exclusively, and who still remain here and there a picturesque element of it. Nature will not be divorced from her children. In their rudest estate their presence enhances her charms, and valley, hill, and stream derive added interest from human association. After many hours, perhaps days, of lonely travel amid wild scenery, when the solitude of forests or the monotonous expanse of great plains has become oppressive, what a relief it was, in the days of youthful adventure, to see the smoke of an Indian camp curling up from a piney gorge, to come suddenly upon the comical bark shelters which served the red man in the higher mountains, or to meet at the turn of a valley stream a village of earth mounds, whose simple denizens regarded the stranger with naïve curiosity. To this day, the few survivors of once numerous tribes remain as picturesque figures in many a landscape that would be less effective without them. They harmonize with earth, and rock, and tree, as well as the larks and quails, and places dispeopled of them seem to lack completeness. How much everywhere the presence of man modifies the aspect of a country. By what he does, or by what he leaves undone, the region he inhabits is made more or less attractive. One race, or one stage of culture, adapting itself to improving upon natural conditions, enhances the beauty of its habitat. Another race or stage of culture, violating or neglecting those conditions, lessens or obliterates that beauty. All the Indian tribes of America lived in such a way as to leave the natural charms of their land unimpaired. They neither extirpated forests nor impoverished the soil. The sites of their encampments and villages were usually the most lovely spots. Even the aborigines of California, reckoned among the lowest of their kind, seemed to have a preference for the prettiest places. In the valleys, their villages would occupy a knoll or bluff overlooking the river and giving far vistas of the flowery plain through natural parks of oak. In the foot-hills, they would be found on some grassy slope by running water or perennial springs, under or near the shelter of pleasant groves. It is common to attribute the selection of such sites to an instinct for the beautiful, but there is really no good reason to credit these Indians, if any, with such a decided feeling for natural beauty as would be required to determine their choice of localities for camps or homes. The ideas and sentiments which make men fond of fine landscapes are largely the result of culture. It is only in the literature of refined nations that they assert themselves. That the Indian village has a fine site or commands the best view in the neighborhood is only a coincidence. He camps or builds where he finds the most suitable conditions for his mode of life. If in the great valleys of California he dwells by stream and grove, it is to be near water and fishing, and where, in the summer, he can be screened from the intense heat of the sun. If he choose a knoll or bluff, it is because his hut and his ricks of acorns and cereals will be secure against the floods that often spread over the lower land. On the same principle he selects banks of brooks or the grassy mounds of springs, in the hills, because they furnish him water and umbrageous shelter. In short, utility and not beauty is his aim; and it happens that just the conditions which are useful to him enhance scenical beauty. By the reflex action of this there may in time be developed in savage man æsthetic appreciation, which doubtless grew at first by some such process of evolution. Without being too curious on this point, however, let us be thankful to whatever cause put the figure of the red man amid scenes that would be less interesting without him. Looking back twenty-five or thirty years, we recall those primitive landscapes where he moved about free and lithe as the antelope, over valleys where wild oats and flowers bordered his trail, where the only architecture was his village of conical earth huts, among which stood poles decorated with the stuffed geese used for decoys to entrap the living bird. Over broad, level areas in the distance we can see a million wild fowl "feeding like one," the aggregate movement of their heads giving a peculiar ripple to the white surface of the vast flock. Herds of elk wend their way to drink from the river, and in the coverts of oak the deer gaze with innocent eyes, unsuspicious of danger. When the Mexican settlers came, the Indians still remained; but herds of domestic cattle disputed the pasture with their remote kindred, the elk; and the flat, oblong houses of sunburnt bricks, with tiled roofs and court-yards, and here and there the towers of a church, gave a character quite Spanish and sophisticated to the scenery. Men change, and nature with them.

It must be said now that the aborigines of California are rapidly passing away. Their number in 1823 was estimated at one hundred thousand. Forty years later, in 1863, returns to the Indian Bureau made it twenty-nine thousand three hundred. At the present time it probably falls below twenty thousand, a quarter part of whom are in government reservations or living under the protection and in the care of farmers; while here and there, especially in the mountains, a few depleted tribes still enjoy the freedom of their ancestors. Many beautiful valleys, once populous with them, know them no longer. The pioneer Yount, who settled in Napa Valley in 1830, used to say that it then contained thousands of Indians of the larger tribe that gave the valley its name; there are now only a few vagabond survivors haunting the purlieus of town and farm. Probably the largest portion of the California Indians was always to be found in the big valleys of the interior, and those lesser ones lying between the spurs of the Coast Range, for it was in these localities that game, fish, seeds, and esculents were most abundant and easily obtained. These lowland tribes looked sleek and well fed. They were more amiable and less warlike than their brethren of the highlands, who were often a terror to them. In the southern part of the State, along the coast, they were largely brought under the influence of the Mission Fathers; but in the northern interior they had not had much contact with our race until after the American occupation. Twenty-one missions were established between the years 1769 and 1820, extending from San Diego, in the extreme southern portion of the State, to the neighborhood of the Bay of San Francisco, near its centre, that at Sonoma being the last and most northerly. In the region above Sonoma, reaching to the Oregon border, and embracing an area three hundred and fifty miles long by one hundred or more wide, the aborigines knew very little of the greedy whites who have since displaced and nearly exterminated them. A few trappers and hunters, mostly Canadians in the employ of the Hudson Bay Company, had visited the head of the Sacramento Valley in search of otter and beaver. Some of the *voyageurs* had even been accompanied by small bands of Oregon Indians, of more nomadic and warlike habits. At one time, early in this century, the Spanish governor of the then province of Alta California sent a military expedition from Monterey to the Sacramento Valley, to drive out some Russians who were reported to be there, but who were not found; and old Gilroy, who was one of the party, used to tell how numerous the Indians were, and how much they were frightened by the discharge of a small howitzer from a mule's back, — for such was the primitive artillery of this quaint expedition. Between 1835 and 1848, American emigrants began to establish "ranches" in the Sacramento and northern coast valleys. The docile natives readily gathered about them, sometimes for protection against the mountain Indians, and even engaged in their service as farm hands. Their labor was always voluntary, and the control over them was usually gentle. No concerted efforts were made to teach them religion or letters. They maintained their tribal organization as before, and followed all their old habits. They were attracted to labor only by their desire for beads, blankets, garments, and some articles of our food of which they became very fond. Gradually, as the lands along the rivers were occupied, game driven away, and their fish-dams torn down to make way for steamboats and sailing craft, the Indians mostly retired to the hills, whence, impelled by the sharp edge of a new appetite, they made thieving descents on cattle-folds and stables. The settlers then too often regarded and treated them as enemies to be killed on sight. Many of the early border-men, who recognized no difference between these and the fierce, more aggressive savages they had known elsewhere, regarded them as natural enemies from the first, and would fire upon them as readily as upon a coyote. As late as 1850, however, many of the northern tribes were living undisturbed in their primitive condition, snaring geese and brant on the plains; crawling upon the antelope in the tall grass with deceiving antlers on their heads; catching salmon and sturgeon nearly as long as themselves; making baskets and network, bows and arrows, and capes of feathers; wearing little clothing, ordinarily, but decorated sticks through the lobes of their ears and fringed aprons of tule (a kind of reed) about their loins; tooting mournfully on their little flutes, made by removing the pith of certain woods; gambling at their native games with excited vociferation; sweating themselves in the great medicine houses; howling over their burnt or buried dead, — for they both cremated and inhumed, — and generally behaving in a way most uncivilized, but quite satisfactory to themselves; a good-natured and harmless race, as a rule, liking the neighborhood of the whites when justly treated, and seldom presuming upon kindness. By their labor on the farms, when most of the whites were digging for gold, they helped in the first development of home agriculture, and thus played an important part in the early resources of the State, as before they had aided in building up and maintaining the mission settlements. It has been only since their numbers were thinned and scattered that anybody has tried to make a thorough, systematic study of their tribal organization, nomenclature, myths, and customs. Stephen Powers has latterly been devoting himself to this useful task with much zeal and success, and when he shall have published together the papers on this subject which have appeared separately in the "Overland Monthly," the public will find the book one of the most interesting contributions to Indianology and what Tylor calls the science of primitive culture.

During the various rambles which furnished the material for these sketches of California scenery the writer was much interested in observing the evidences of former Indian occupation and handicraft. He had seen, a quarter of a century ago, that the tribes unaffected by contact with our civilization presented a perfect picture of the arts and customs of the later Stone age, when implements or weapons were polished, and when woven and braided fabrics were made, and earthen huts gave the first kind of architecture. He had exhumed from considerable depths in the auriferous gravel deposits of the Sierra stone mortars and pestles and arrow-heads, like those still used by living tribes. In later journeys, therefore, it was a pleasant incidental task to follow again in the footsteps of the first people. There is no reason to believe that any tribes dwelt permanently at great elevations in the Sierra Nevada, if anywhere within the deep snow-line. In the summit valleys, about the lakes, and at the sources of streams, where these wild children of nature would find it most convenient and pleasant to live, the elevation above the sea is from five thousand to seven thousand feet, and the snow falls to a depth of from ten to twenty feet, continuing on the ground from November or December until June or July. Most of the lakes at this season are frozen and covered with snow; even the smaller streams are often banked over with snow; and the game has fled to the lower portion of the range. But while the Sierra was not the constant home of the Indians, they resorted thither regularly in the summer season, from June or July to November, except when they were denizens of the great lower valleys, which supplied them with all they needed in every season; these were, moreover, occupied by the less warlike tribes, who were seldom able to cope with their hereditary foemen of the mountains. The summit region of the Sierra Nevada furnished good fishing in its lakes and some of its streams; deer and mountain quail and grouse abounded; huckleberries, thimble-berries, wild plums, choke-cherries, gooseberries, and various edible roots were tolerably plentiful; the furry marten, weasel-like animals, woodchucks, and squirrels were tempting prey; the water was better, and the climate cooler, than at a less elevation; hence this region was the summer resort of Indians from both slopes of the range, and often the possession of a valley by lake or river was decided by battle between the various tribes from Nevada and California. The Hetch-Hetchy Valley, or "Little Yosemite," for instance, was, up to a very recent date, disputed ground between the Pah-Utahs, from the eastern slope, and the Big Creek Indians, from the western slope, who had several fights, in which the Pah-Utahs (commonly called Piutes) were victorious. This statement was made to the California Academy of Sciences by Mr. C. F. Hoffman of the State Geological Survey, on the authority of Joseph Screech, a mountaineer of that region; and similar statements have been made to the writer by old mountaineers, with reference to the Yosemite Valley and other former aboriginal resorts along the summit of the Sierra. As the mountain Indians, and those of the Nevada plateau, were comparatively nomadic in their habits, they left few or none of the large black mounds, indicating long and constant residence, which were left so abundantly by the mud-hut builders of the Sacramento basin. Pieces of bark stripped from fallen pines or firs, and slanted on end against tree-trunks or poles, with a circle of stones in front for a fire-place, were the usual shelter of the California mountain tribes, except that in the northern extremity of the State, where the winter climate is more rigorous, some of the tribes — notably the Klamaths and their congeners — built huts of roughly hewn logs, employing bark and brush shelters only in their summer fishing and hunting excursions. Speaking generally, therefore, the mountain Indians have left few traces of themselves, except the stone implements which are occasionally unearthed, or still found in the possession of the wretched remnants of once powerful tribes.

Along the summit of the Sierra Nevada there is scarcely any memento of them to be found, except the arrow-heads shot away in hunting or fighting, or the broken arrow-heads and chips from the same to be gathered at places which have evidently been factories of aboriginal weapons. The most notable find of this latter sort made by the writer was at the Summit Soda Springs, a most picturesque spot at the head of the northernmost fork of the American River, nine miles south of Summit Valley Station, on the Central Pacific Railroad. Here, at an elevation of about six thousand three hundred feet above the sea, the river breaks through a tremendous exposure of granite, which it has worn into natural gorges several hundred feet deep, except where it runs rapidly through valley-like glades of coniferous woods, in which the new soil is covered with a rank growth of grasses, flowering plants, and shrubs, where the deer come to drink at the salt-licks, and the piping of quails is constantly heard, alternating with the scolding cry of jays and the not unpleasant caw of the white-spotted Clark crow. Just in the rear of the public house kept at this locality, the river tumbles in slight falls and cascades over slanting or perpendicular walls of richly colored granite, shaded by beautiful groves of cedar and yellow pine, which grow in the clefts of the rock to the very edge of the stream, and crown the dark cliffs above. On the rounded tops of the ledge overlooking these foaming waters, on both sides of the stream, the Indians used to sit fashioning arrow-heads and other weapons of stone. This was their rude but romantic workshop; and the evidences of their trade are abundant on the sloping rock, in the coarse, granitic soil which forms the talus of the ledge, and in the blackened litter of their ancient camp-fires. They have left one record of themselves at this locality which is quite remarkable. A shelving ledge of granite on the right bank of the stream, worn to an even and almost smooth surface by glaciers or snow-slides, is covered for a hundred feet with rudely scratched characters, circles or shields inclosing what may have been meant for animal forms or other symbols of expression. They appear to have been cut or scratched on the ledge in comparatively recent times, for the very shallow incisions reveal a fresher rock than the general surface. The California Indians are not known to have possessed any method of writing, pictorial or otherwise; but these curious rock markings may have had some meaning to the people who made them. In the *débris* about this sculptured ledge, as well as in that among the rocks on the other side of the river, before it had been disturbed by visitors to the springs, fragments of arrow-heads, and chips of the materials composing them, could readily be found. Their flat shape and light specific gravity caused them to wash to the top and one had only to look carefully, lightly raking with finger or stick the superficial gravel, to find many curious specimens. In this peculiar quest many persons, who cared nothing for the scientific or artistic suggestions of the simple objects sought, developed a strong interest. It kept them out of doors with nature; it gave them a pretext for remaining in the air by a lovely scene; it aroused that subtle sympathy which is excited in all but the dullest minds by the evidences of human association with inanimate things, and particularly by the relics of a race and a life which belong to the past.

The Indians that congregated at this point, summer after summer, whether from Utah or California, employed in arrow-head making every variety of siliceous rock, of slate, spar, and obsidian or volcanic glass. The larger heads were made of slate and obsidian, which materials also served for spear-heads, used in spearing fish, and from two to four inches long. Obsidian seems to have been better adapted for all sorts of heads than any other material. It could be shaped with less risk of breaking in the process, and could be chipped to a much sharper edge and point. The points of some of the small obsidian heads gathered by the writer are so keen, even after burial or surface floating, that a slight pressure will drive them into the skin of the finger. The greater number of small arrow-heads found, as well as the greater proportion of the chips, consisted of jasper and agate, variously and beautifully colored and marked; of obsidian, of chalcedony, of smoky quartz and feldspar; very rarely of quartz crystal, and in only one instance of carnelian. While the larger heads measure from an inch and a half to four inches in length, with a breadth of half an inch to an inch and a half in the widest part, the smaller heads measure only from three quarters of an inch to an inch in length, their greatest breadth being seldom more than half an inch. The latter were evidently intended for small game, such as birds and squirrels. The workmen seem to have had more difficulty in making them, for they are often found broken and imperfect.

This was due, not only to their size, but also chiefly to the difference in material, when the small vein-rocks were used, these breaking with a less even fracture, and being full of flaws. Persistence in the use of such uncertain material, when obsidian was so much better adapted to the purpose and equally abundant, would seem to have been dictated by a rudimental taste for the beautiful. A collection of the jasper, chalcedony, agate, and crystal heads and chips presents a very pretty mixture of colors, and the tints and handsome markings of these rocks could not but have influenced their selection by the Indians, who spent upon their manipulation an infinite amount of care and patience. It is interesting to note even so slight an evidence of taste in these savages of the Sierra, especially when we remember it was supplemented by the artistic finish they gave to their bows and to the feathered shaft that bore the arrow-head, no less than to the quiver of wild skin in which the arrows were carried. There is some reason to suppose that the selection of the above materials may occasionally have been decided by the superstitious attribution to them of occult qualities. Nearly all aboriginal tribes, and even some civilized races, have attached a peculiar sanctity and potency to certain stones, and the Chinese to this day give a religious significance to jade. It is uncertain, however, to what extent such notions obtained among, and influenced the simple savages of California. None of the rocks used at this Indian workshop were obtained in the locality. The writer was able to trace their origin to Lake Tahoe, across the western crest of the Sierra, and not less than twelve or fifteen miles from the Soda Springs by any possible trail. There they are so abundant as to have partly formed the beautiful gravel beach for which the lake is so famous. The obsidian came from the ancient craters that adjoin the lake, the source of those enormous ridges of volcanic material which form its outlet, the cañon of Truckee River. Doubtless the flints, slates, and obsidian of this region formed objects of barter with the lower country Indians; for the writer remembers seeing arrow-heads of such material among the Sacramento Valley tribes twenty-five years ago. On the Lake Tahoe beaches are sometimes found spear-heads five inches long, with perhaps an inch of their original length broken off, generally at the barbed end. Similar materials were used and to some extent are still used by the mountain Indians in the northern Sierra as far as Mount Shasta, the rocks of the crest furnishing them everywhere along the line of volcanic peaks which dominate the range. In the Coast Range supplies of obsidian were obtained by the northern tribes from the region about Clear Lake, where there is an entire mountain of this material. The antiquity and former great number of the tribes in this region are attested by the wash of obsidian arrow and spear-heads, flakes and chips, about the shore of the lake. The beach at the lower end is fairly shingled with them. About the flanks of Mount Shasta, especially on the McCloud River side, obsidian is again very plentiful, and, with some beautifully variegated jaspers, seems to have been most used. The writer found extensive chippings of it at several points on the head-waters of the Sacramento, notably at Bailey's Soda Springs, thirteen miles south of Strawberry Valley, where the Castle Rocks — fantastic crags of granite — push up through the slates and lavas of the neighborhood two thousand five hundred feet above the river. Here, as at the Summit Soda Springs, the Indians had chosen one of the most charmingly picturesque spots for an arrow-head factory. But here, as elsewhere, something else than an instinct for the beautiful moved them in their choice of locality. There is fine trout and salmon fishing in the river, while there are no fish at all in the upper North American, near the Summit Springs, owing to the falls which prevent fish from ascending.

Again, the snow-fall is not so great on the Sacramento as to drive the Indians away in the winter. Its banks are their preferred home at all seasons. There they still fish and hunt, and are more nearly in a primitive condition than their kindred farther south, who are now few in number and more or less domesticated with the whites. Since the Indians of the Sierra Nevada came into familiar contact with the whites, they have adopted fire-arms in preference to bows and arrows, when they can obtain them, and, where they retain the latter, now generally use metal or artificial glass in making arrow or spear heads. In a great measure, also, they have abandoned the use of the stone-mortars employed for so many years by their ancestors, and which about Mount Shasta, as perhaps in other volcanic regions, were made of trachyte, as certain other implements were made of red lava.

It may increase the interest of this sketch to describe the method used in the manufacture of arrow-heads, which was the first trade of primitive man. Mr. E. G. Waite, in a paper contributed to the "Overland Monthly," described as follows the process he saw in use among the Indians of central and northern California, in the early days of American settlement. The rock of flint or obsidian, esteemed by the natives for arrow-pointing, is first broken into flat pieces, and then wrought into shape after this fashion: "The palm of the left hand is covered by buckskin, held in its place by the thumb being thrust through a hole in it. The inchoate arrow-head is laid on this pad along the thick of the thumb, the points of the fingers pressing it firmly down. The instrument used to shape the stone is the end of a deer's antler, from four to six inches in length, held in the right hand. The small round point of this is judiciously pressed upward on the edge of the stone, cleaving it away underward in small scales. The arrow-head is frequently turned around and over to cleave away as much from one side as the other, and give it the desired size and shape. It is a work of no little care and skill to make even so rude an instrument as an arrow-head seems to be, only the most expert being successful at the business. Old men are usually seen at this employment. This manufacture of arrow-heads by a primitive people readily suggests the origin of trade. In the earlier stages of human development, when man wore a skull of the Neanderthal type, the maker of the best weapons was the most successful in coping with the cave bear, hyena, and other animals of the period. His arrowheads and other arms of stone were models. Superstition invested them also with an infallible gift to kill. His well shaped and charmed weapons would be sought after. Suppose this ancient troglodyte and mighty Nimrod should be wounded and crippled for life in one of his fierce encounters with formidable beasts, what would self-preservation demand, what would be the unanimous voice of his tribe, but that he must become the armorer for the whole? What better could he do than fashion the arms that would furnish the most food for himself, his family, his kind? 'Bring me, then,' he would say, 'a certain share of the fruits of the chase, and I will give you the instruments that yield the surest rewards.' Here, then, a skilled artisan began his workshop, the chips of which in piles survive him by thousands of years in the caves of the old world. Thus barter began, and man, little by little engaged in diversity of employment, according to natural or acquired abilities." The method of finishing arrow-heads described by Mr. Waite as prevailing among the California Indians is substantially the same as that observed by A. W. Chase of the United States Coast Survey, among the Klamaths so recently as 1873. A drawing made by him of the implement used by the artisan of this tribe closely resembles the figure of such an implement given in Tylor's work on prehistoric art in Europe. Catlin describes a similar method and instrument in use among the tribes east of the Rocky Mountains. They broke a cobble of flint with a rounded pebble of horn-stone set in a twisted withe as a handle, then selected such pieces as from the angles of their fracture and thickness would answer as a basis of an arrow-head. The finishing process is described as follows: "The master workman, seated on the ground, lays one of these flakes on the palm of his left hand, holding it firmly down with two or more fingers of the same hand, and with his right hand, between the thumb and two forefingers, place his chisel or punch (made of bone) on the point that is to be broken off; and a coöperator (a striker) sitting in front of him, with a mallet of very hard wood, strikes the chisel or punch on the upper end, flaking the flint off on the under side, below each projecting point that is struck. The flint is turned and chipped until the required shape and dimensions are obtained, all the fractures being made on the palm of the hand." This is more elaborate than the California method, which was carried on by a single workman. Catlin also describes his two artisans as singing exactly in time with the strokes of the mallet. Leaving out the minor differences, there is a strong likeness in all the primitive methods and implements the world over, showing the instinctive readiness of the race to originate independently the same methods and forms under the same circumstances.

Going back to the days before the pale-face invaded their land, one can easily recall groups of California aborigines, seated on the picturesque lake and river spots chosen for their homes or summer resorts, sorting out the beautiful stones they had procured for arrow-heads, and chipping away slowly as they chatted and laughed, while the river sang, or the cataract brawled, or the pine woods soughed, as musically and kindly to them as to us.

SONG OF THE VAQUERO.

A LIFE on the prairie long and wide,
Where the wild oats roll in golden tide,
And the hills are blue on either side.

A life on the fleet and eager steed,
With strain of Arab in his breed,
Circling around the herd as they feed.

A life as free as the air I drink,
That flows like wine from the bubbling brink
Of glasses that touch in social drink.

Ha! With a toss of my lasso true,
The stoutest bull of the herd I threw,
As over the vale he wildly flew.

Ha! When the grizzly ventured afield,
Leaving the shelter chaparrals yield,
He fell a prey to the loop I wield.

Ha! With a skill that was surer yet,
I flung my terrible lariat,
And dragged to his death the foe I met.

Juanita smiles as I gallop by;
Soft is the light of her darkling eye,
And red is her lip as the berry's dye.

Juanita smiles, for she knows the hand
That flies the lasso, and the marking brand
With equal skill can the lute command.

Juanita smiles, for she knows the time
Is fixed, for the Mission's wedding chime
When the rain has brought the flowers prime.

Then the glad festa's joy will begin,
The castanet and guitar's sweet din,
As the neighbors all come trooping in.

Then will the dancers happily beat
The waltz of Castile with lightsome feet,
While horsemen race in the contest fleet.

Then life will be sweet to groom and bride,
Where the wild oats roll their golden tide,
And the hills are blue on either side.

THE TRINITY DIAMOND.

It was a hot June day in 1850, when we started, Brandy and I, from the American River, where we had been for nearly a year unsuccessfully mining, to seek our fortunes on the Trinity. A tramp of three hundred miles, through a lonely valley and over rugged mountains, lay before us; but we were full of pluck and strength. Glowing reports had reached us from the far north, and we liked adventure. The country was new, strange, and unpeopled. It seemed as foreign to us as the West Indies and Mexico did to the Spanish adventurers under Columbus and Cortez, and we had the same golden dreams that lured those pioneers, tinging all our future with blissful hopes. Imagine two young fellows, with unkempt locks, under broad-brimmed felt hats of a drab color, clad in gray woolen shirts, and blue dungaree trousers — the latter held up by a leather belt about the waist, and tucked into long-legged boots, the belt itself holding a sheath-knife, revolver, tin drinking-cup, and rubber flask; on their backs neatly bundled blankets, strapped across their shoulders, and inclosing a small package of raw pork, sea-biscuit, and tea, while over each bundle lay, bottom up, a large tin pan, glistening in the sun, and suggesting visions of the dairy and rural homes far away. There you have the portraits of two prospectors. We belonged to the noble army of explorers that found and opened the treasure-vaults of the Sierra Nevada and Rocky Mountains; that planted the seeds of empire from the upper Missouri to the Pacific; that whitened western seas and streams with the sails of a new commerce, laid an iron road across the continent, and aroused the sluggish civilization of Asia to new motives. Those heroes of the pick and pan were not romantic figures; their triumphs were not bloody ones; but see what they achieved for the world, and cease to despise them if they failed to achieve much for themselves.

As for Brandy and I, we trudged on, chatting, whistling, and singing, intent on finding virgin gold-beds far from the crowded placers we had left. We had read Humboldt; had traced the gold formation through South and Central America and Mexico to California; fancied it must link farther north with that in Siberia, and the Ural Chain, and were resolved to push even beyond the Trinity, if that stream did not enrich us speedily. Our mining implements, a tent, some cooking utensils, a few clothes, and several months' supply of salt meat and flour, we had sent ahead in a wagon to Reading's Springs, in the Shasta Hills, whence they were to be transferred by pack-mules to Trinity River on our arrival. The scanty provisions we carried on our backs we expected to eke out with occasional meals at the *ranchos* along the Sacramento River. One of us carried a rifle, for protection against any unfriendly Indian or savage beast that might obstruct our way. Thus equipped we pushed ahead, averaging thirty miles a day with ease. The level valley was covered with a ripening growth of wild oats, and looked like a vast harvest-field, bounded on one side by the purple wall of the Coast Range, on the other by the hazy outlines of the more distant Sierra, and ahead only by the dazzling sky, save where an occasional grove of oaks marked a bend or branch of the river, and loomed up in the hot, shimmering air, with an effect as if a silvery sheet of water flooded its site. It was a lovely spectacle, as this sea of grain, in places as high as our heads, waved its yellowing surface like a true ocean. The road through it was not well defined after we left Knight's Landing, and we wandered off by Indian trails far from the river; so that, on one occasion, we traveled sixty miles before meeting with water fit to drink. A few pools, the remnants of the previous winter's flood, were found in hollows of blue, clayey soil, hot, putrescent, and sickening. At one such place, where a lone tree broke the monotony of the plain, the air was populous with dragon-flies of great size and brilliant colors, whose gauzy wings often touched our hands and faces, while swarms of yellow hornets hovered over the mud and myriads of mosquitoes hummed their maddening song. A few yellow blossoms still flaunted their beauty on the spot, though most of the plants had been trampled down by thirsty cattle. We pushed on till late in the night, then spread our blankets on the earth, and, regardless of the coyotes that barked querulously around us, slept under a roof of splendid stars.

What a delight it was, after a hot tramp, to reach a clear, pebbly creek, to drink and bathe in its waters, and then, under a grove of noble oaks trellised with vines, to drink from the adjoining *rancho*, and eat blackberries picked by the Indians along the stream. At that time the settlements on the upper Sacramento were few and far between. They consisted of an adobe house or two, tenanted by a family of mixed races, the man being an American or European, the woman a Mexican or Kanaka; while near by were the earthen huts of a few amiable Digger Indians, who did the fishing and hunting, and most of the farm-work, satisfied with blankets, coffee and sugar, and a few old clothes, for their wages. These *ranchos* were usually on the bank of the Sacramento or some confluent, and were stocked with large herds of half-wild cattle. Some of them became the sites of towns at a later day. Their owners were very hospitable to the few adventurers who called on them before the grand rush to the northern mines set in, and I often recall their hearty words and homely cheer with gratitude.

One night we stopped at a log cabin lately built by Missouri squatters. As we neared it, some time after dark, we heard the sound of a fiddle, went to the open doorway, and looked in. There was a rude bar garnished with a few black bottles. At one end of the bar sat the fiddler upon a keg, while a number of stout fellows in buckskin were leaning on the bar, or against the log walls. Presently a tall, broad-shouldered man in a butternut suit opened a rough "shake" door leading into a second apartment, and shouted, "Gentlemen, make way for the ladies!" At this he led forth a female who was "fat and forty," but hardly fair, — a very short and plump person, clad in plain calico, her face shining as if it had been oiled, her eyes bright with laughter. Behind her came a thin girl of ten or twelve years, who bore traces of a recent struggle with fever and ague, and whose yellow hair hung down in two big braids, tied with blue ribbons. There was to be a dance, and these were the ladies. The fiddler struck up "The Arkansas Traveler," and the ball began. Of course every gentleman had to wait his turn for a partner, except as they made what were called "stag couples." It must be said that the ladies were compliant and enduring. They danced with everybody and nearly all the time. They even invited the "stranger" at their gate to "take a turn," — an invitation that youthful modesty alone caused us to decline. When we went to sleep under the big oak fronting the cabin, the rasping tones of the backwoods fiddler were still heard, as also the clat-clat of the loose planks on the cabin floor keeping time.

At last we reached Reading's Springs, — a famous mining camp in those days, which has since grown into the town of Shasta. Here we gave the charge of our outfit to the Mexican owners of a pack-train, and started with them across the mountains for Trinity River. The train consisted of about thirty mules; and we helped to drive them over a narrow trail which had been marked out with no regard to easy gradients. The heavily laden brutes grunted and groaned as they tugged up the steep, conical hills between Shasta and Trinity Mountain. They would often run off into the woods, and then the shouts and curses of the Mexicans, although in mellow Spanish, were startling to the very trees and rocks. But the hardships of the trip only gave a keener zest to our enjoyment of the mountain air and water, so delicious after our experience in the valley; of the luxuriant and varied vegetation, the aromatic odors of the pines, the music of rippling brooks, the dizzy glimpses of vaporous cañons yawning below, the noble vistas of far peaks as we climbed higher and higher, and sat with beating hearts and white lips on the summit of Trinity Mountain. Descending this elevation, we reached the river of our hopes, followed its course to the North Fork, and pitched our tent under a tall yellow pine on the bar below the mouth of that stream. Trinity River is a cold nymph of the hills. All its course is through the tumultuous peaks that mark the blending of the northern Sierra and Coast Range; and it has always a touch of its native ice. Whirling through rocky cañons with foam and roar; darkened by overhanging precipices, by interlacing pine and fir, or hanging vines; gliding into narrow valleys, that margin it with meadows and tremulous-leaved cotton-woods, and spreading out in broader bottoms to coax the sun, — it is still the same cold stream, until it reaches the literally golden sands of its ocean outlet through the Klamath. When we saw it in 1850, it was beautifully clear, and its wooded banks were wildly picturesque. Hardly more than fifty miners were trying to tear the golden secret from its breast, and the emptyings from their rockers did not sully its purity. Indians fished in it, and the deadly combats of the male salmon often sent free offerings to their hands. The miners themselves would sometimes watch these finny tragedies, and swim after the vanquished lover for their dinner.

It was a new sensation to strike our picks into the virgin cobble-beds, among tuft grass and thickets of rose-brier; to overturn gray boulders, never disturbed before; to shovel up from the soft bed-rock the gold-seeded gravel that promised a harvest of comfort and happiness. It was pleasant to have our sweating toil eased by the cool breezes that daily blew up from the sea; though when one of these breezes became a gale, tossed the coals from our camp-fire into our poor tent, and lighted a flame that consumed our shelter and supplies, making the rifle and pistols fire an irregular salute, the sea wind was not blessed. The nearest trading-post was ten miles below, at Big Bar; and a weary journey it was, over a lofty mountain, to reach it, while all that we bought had to be packed on our own backs. Beef-cattle were lowered down the steep descent by the aid of ropes, and their flesh was precious. The butcher of Big Flat was an eccentric Yankee. As meat was fifty cents a pound, the portions without bone were in great demand, for economical reasons. Liver was in particular request. As it was impossible to find an ox all liver, and the Strasbourg goose-fattening process would not apply to cattle, our butcher was obliged to adopt some plan to relieve himself of a difficulty. It was his habit, when a customer asked for liver, to inquire, "Have you a canvas-patch where you sit down?" And when the customer would naturally respond, "Why, what's that got to do with it?" he would answer, philosophically, "If you haven't got a patch on your breeches, you can't have any liver; that's what. There isn't liver enough for everybody; there's got to be something to discriminate by, and it might as well be a canvas-patch as anything else." And to this impartial rule he faithfully adhered, albeit canvas-patches began to multiply, and other parts of the animal economy, like the heart, had to be pressed into service.

On the bar where Brandy and I opened a claim and started our rockers only three more men were working. They owned and operated in common a large quicksilver machine. We soon knew them as Peter the Dane, English George, and Missouri. The nomenclature of the early mining epoch was original and descriptive. Individuals, like places, were named in a way to indicate peculiar traits or circumstances. Thus, my partner, Brandy, whose real name was William, — a slender, fair-skinned, blue-eyed fellow, of temperate habits, — had a high color in his cheeks that a rough comrade called a brandy-blush. The joke was too good not to live, and so the name of Brandy clung to him for years, being varied occasionally to Cognac, by way of elegant euphemism. Our Trinity River neighbors were all named from their nativity, the signs of which they bore plainly in speech and looks.

Peter had served in the navies of three nations,

ending with the United States. He was a young man of cultivation and genius; kept a journal in Greek, to conceal its secrets from his comrades before the mast; acquired English from the library of the man-of-war Ohio; had a good knowledge of our literature; spoke French and German well; was a clever draftsman and musician, and a witty, brilliant talker. But he was only Peter the Dane, except, indeed, when called "Dutch Pete" by a class of Americans who think everybody Dutch (or German) who says "ja." We sympathized on the subject of poetry and music. Indeed, it was my whistling "Casta Diva," while rocking the cradle, that made us acquainted. He used to recite poems from the Danish of Oehlenschläger, which I would render into English verse. He went through "Hakon Jarl" in that way — the recitation at night by our camp-fire, the pines soughing overhead, the river roaring below; a truly appropriate scene for a Norse epic. The ink to write out my translation I made from the juice of ripe elder berries. One night Peter and I went to Big Bar, and crossed the river by crawling over the branched top of an Indian fish-dam, on our hands and knees, to hear a violin that somebody owned in that wild place. The night was so pitchy dark that we could not see the white foam on the rapids around us; and we did not know what a fool-hardy feat we had performed till the next day.

George was a simple-minded, ignorant Englishman, credulous and kind-hearted, who had made a voyage or two, when he heard of the gold discovery, worked his passage to San Francisco, and had drifted up to the Trinity, in eager quest of a fortune for his old parents and his sweetheart in England. He was a good worker, and a good listener. It was curious that two such men should come together; more curious they should have for a partner Missouri, — familiarly called "Misery," — a lank, sallow man, with long, straight, yellow hair, tobacco-oozing mouth, broad Western speech, a habit of exaggeration that was always astonishing, and a cold selfishness that he took no pains to conceal. My partner, Brandy, had been a dentist in New York, was still ready to pull or fill a tooth, and enjoyed as much as others the tones of his rich baritone voice in laugh or song.

These comprised the company that used to meet about a common fire at night, smoke their pipes together, talk of home and its friends, exchange experiences, tell stories, sing songs, and crack jokes at one another's expense. Peter used to tell of his adventures at sea; often with so much humor that we laughed till our sides ached. "Misery" related his adventures with "bars" and "Injins," and told us how he "made things bile" when he mined at Hangtown, where the gold poured down his "Tom" in "a yaller stream." Brandy used to sing "The Old Folks at Home," until the tears came into all eyes but Missouri's though even he grew quiet under its influence. From how many thousand mining-camps, in early years, — before daily mails, telegraphs, and Pacific Railroad, — went up that song of the heart, with its tender, refining, and saving influence! Well might old Fletcher say, "Let me make the songs of a nation, and I care not who makes its laws."

Sometimes we got into controversies — not on politics, for we never saw newspapers nor heard politicians; nor on religion, for we did not know certainly what day was Sunday, nor care for creeds, so long as men were honest and kind. But literary memories, and subjects connected with our daily life, would provoke talk enough. One night I wondered if there might not be diamonds in the gold deposits of California — why not along Trinity River? I had found some very small rubies.

"Oh," said Peter, "they are likely enough to be found, if we would only look for them. I have fancied them rolling off the hopper of our machine many a time. They have been found in the mines of the Ural, and I was even told of small ones being found in the southern dry diggings of California. You know something about precious stones, Brandy, what do you think?" Brandy rejoined: "It is true the diamond is found in gold formations, associated with clay or drift, as in Brazil, Georgia, and North Carolina. The most famous district is Golconda,

Hindostan. In the rough, the stone looks like a quartz pebble, or one of the bits of rounded glass found on sea-beaches near cities. Unless a person was familiar with its appearance in this state, he would surely throw it away as worthless. If it was fractured and of good size it might attract attention by its lustre, and be saved by one ignorant of its real nature as a pretty stone."

George listened to this speech with unusual interest. Missouri declared his intention to look out for ground pebbles "mighty sharp" after this.

Brandy added that diamonds were sometimes found in connection with oxide of iron, and might have a metallic look on their rough surface; and at this George gave him a quick, keen glance.

"Well, it would pay better to find a big diamond than a gold-mine," said Peter. "Napoleon had a single diamond in the hilt of his sword of state that was worth a million dollars. It weighed four hundred and ten carats. The Braganza diamond weighs sixteen hundred and eighty grains, and is valued at twenty-eight million dollars."

Here "Misery" gave a long whistle, followed by a yelping laugh, and the characteristic exclamation, "That takes my pile."

"How big are diamonds found?" asked George, after the laughter excited by the Missourian's racy expression of incredulity had subsided.

"Oh, half the size of an egg; as big as a walnut, sometimes," said Brandy, rather wildly.

"As big as a piece of chalk," added "Misery," with a leer that let out the tobacco-juice.

Peter remembered that Empress Catherine of Russia bought of a Greek merchant a diamond as large as a pigeon's egg, which had formed the eye of an idol in India. A French soldier stole it from the pagoda, and sold it for a trifle. ("What a dumb fool!" interposed Missouri.) The Greek got four hundred and fifty thousand dollars for it, an annuity of twenty thousand dollars, and a title of nobility.

George's eyes dilated. I had never seen him taking so much interest in any conversation.

"Ah! if we could only find the other eye?" I suggested, "we might all quit this slavish work."

"That reminds me," struck in Peter, "that it is the custom in Brazil to liberate a negro who finds a diamond of over seventeen and a half carats. The search there is followed by some thousands of slaves digging like us. Since we must dig anyhow, why not keep a keen eye on the hopper?"

"Wall, Brandy, kin yer tell us how the diamond comes?" asked Missouri.

"I guess they grow," replied Brandy, with a merry laugh, and a wink at me.

"Perhaps there is more in that than you think," said Peter. "The diamond is proved to have minute cavities; and as it was formed from a solution, it must have been once in a soft state. It may enlarge when left in its original place — eh? The darkies believe that diamonds grow; and perhaps this notion originated from their being found sometimes in clusters, like crystals of quartz. The natives of Golconda had the same notion formerly. They felt for the diamond with their naked feet, in a black clay, as we hunted for clams at low tide in happy valley, boys."

All laughed at this conceit. My partner thought Peter was joking altogether. The latter said gravely he could quote good authority.

"I remember when I was on board the Ohio, reading the travels of Sir John Mandeville. He relates that in Ethiopia the diamonds were as large as beans or hazel-nuts, square, and pointed on all sides without artificial working, growing together, male and female, nourished by the dew of heaven, and bringing forth children that multiply and grow all the year. He testifies that he knew from experiment that if a man kept a small one and wet it with May-dew often, it would grow annually, and wax great."

Here there was another laugh, in which Peter joined. George alone looked serious, and inquired if diamonds might not be even bigger than any that had been mentioned. Brandy thought they might be; he knew nothing to prevent it. A diamond was of no more account in Nature's operations than any other stone.

George then related, with nervous haste, his native

dialect coming out strong as he spoke, that when he was alone on the river, before he went down to San Francisco for supplies, he found a curious-looking stone in the hopper of his cradle. As he was rubbing down some lumps of clay with one hand, while he poured water from a dipper with the other, this stone became very clear, and seemed to have a glazed, metallic coating, except on the side where it had been broken. He picked it out and threw it on the dry sand behind him, intending to take it to his tent "jest for fun loike." A few minutes later, as he sat rocking again, his eye fell on the stone where it lay, dry, fractured side up, "flashin' in the sun jest loike a dimon', but colored loike a rainbow." He thought it a pretty thing to keep, saved it, and when he went to the Bay took it along with him, and left it in the locker of his sea-chest, at a miner's boarding-house on Pacific Wharf. "An' noo I wonder," he continued, almost breathlessly, "if it were na a dimon' truly."

Missouri — who was in the habit of gibing George,

as one ignorant man will often gibe another more simple than himself — did not laugh at him, nor utter contemptuous comment. He sat eying him in attentive silence, with the look that I fancied he may have worn when he "turned up the belly of an Ingin on the creek," as he had boasted one day he did. He had lived in Oregon years before, and "thought no more of shootin' one o' them red devils than a rattlesnake."

Peter asked how big the stone was, and George replied that it was as big as his fist. Brandy suggested it was a fine quartz crystal. If it were a diamond, it would be worth more than anybody could afford to pay; and George might have to remain poor after all, for want of a purchaser.

Peter gravely observed, there was no reason why a larger diamond than any yet known might not be found on Trinity River. As they had not found much gold, there was more room for precious stones, and a big one could be divided and sold easily enough.

George said that it was very bright. He had often seen it shining in his tent at night; and when he put it in the till of his chest, it shone there in the dark. He declared he meant to show it to a jeweler, when he went down again. It might be worth "somethink," if it were no diamond.

Missouri still listened in silence; and no more was said on the subject by any one. Brandy stirred up the embers of the fire, we lit our pipes again, smoked a short time, sang "The Old Folks at Home," and, separating for the night, went each one to his blankets and to sleep, while the wind roared through the pines like a beating surf, and the rapids rumbled and thundered through the rocky cañon of the river.

The next day Missouri said he was going up the river, to a new trading station he had learned was recently started there, to get some tobacco and powder. As he might stay over night he would take his blankets, and his rifle of course, for that he always carried on his shortest excursions. He insisted on a division of the amalgam, as he always did when going to the store, because he was an inveterate gambler at poker, and every store had then its gambling table. His partners had long since learned that it was useless to remonstrate with him; so they weighed him his dust, gave him two or three commissions, and off he went, whistling "The Arkansas Traveler."

We never saw him again. Days passed without tidings of him. We thought he must either have fallen a victim to a grizzly, or to one of his old enemies, the Indians. One of us went up to the new store at last, and learned that he had not stopped there, except for a drink of whiskey, but had pushed across the mountains toward Weaverville, on the road to Shasta. His abrupt departure excited little speculation, and was then passed over by all except George, who referred to it at intervals, and became unaccountably moody and discontented. One night he said he had made up his mind to go to San Francisco: he was sure there must be letters from England. Peter tried to dissuade him from leaving, and told him he could send for letters by express from Weaverville, at a cost of a few dollars. No; he would go. Those express fellows never got anything. Besides, he was "tired of these diggings." He sold his share in the quicksilver machine, took his gold and blankets, and started off, after a hearty hand-shaking from each of the three men he left. We all liked the simple-hearted fellow, and were sorry to see him go; but as we had determined to prospect the streams toward the Oregon line, which had not then been proved to contain gold, we would not pull up and go with him. He promised to send us word if he found good mines after his visit to the Bay, and told us where he would stop while there, — at a house on Pacific Wharf, much frequented by sailors and miners, where he had left his chest. Peter laughingly told him to be sure and get a good price for his diamond; but he did not laugh in reply. Uttering only some kindly words, he wrung our hands again, and we saw him disappear in the woods up the hill.

Later in the summer we prospected several of the

northern streams, finding gold everywhere. But the Indians were threatening; there were no trading-posts; there was not time to get supplies of our own from Sacramento before winter would set in; and at last we all concluded to return to the lower part of the State. I went as far as San Francisco, and the next day after my arrival visited the place on Pacific Wharf, described by George, to inquire after him. It was a thin shell of a house, erected at one side of the wharf on the hulk of a bark, that after years of brave service on the ocean had been sunk and abandoned at last in the dock mud. Only a year old, this house yet had the appearance of age, so weather-stained and toppling was it. Its lower story was divided into a rude bar-room, eating-hall, and kitchen. Its upper floor was covered with what the sailors call "standee berths," provided only with a straw mattress, pillow, and a pair of heavy, dirty blankets. Under many of these berths sea-chests had been left on storage by their owners, mostly sailors, who had deserted their ships to run off to the mines. The landlord himself was an "old salt"—an Englishman. I asked him if he knew George, and could tell me what had become of him.

"Be you a friend of the lad?" he inquired.

I assured him that I was — that we had worked by one another on Trinity River, and he had promised to write to me.

"Well, it's a queer story," said the landlord, a short, thick-built man, with ruddy face, who spoke his mother tongue with many elisions. "Ye see, George came rushin' in one night from the steamer McKim — she as runs 'tween here an' Sacramento. He was down from the mines, he said, an' 'ad come to see ole friends and take away his traps. I told him he would find all there in the chest — ye can see it under the bar 'ere yet, sir — all that his friend had n't taken away. 'Taken away — friend — what friend?' said George. 'Why, your friend from Trinity River,' sez I; 'the feller with the long, tow hair and fever an' ager face, and with terbacker-juice runnin' out of his mouth.' 'Has — he — been — here?' sez George, slow like. 'Yes, he's been here,' sez I, and tell I you sent him for some little things in the chest. 'There it is,' sez I; and after he had treated like a gentleman, he pulled it out, took somethink from the till, put it in his shirt pocket, and went off. Before I could tell him more, sir, the lad — George, sir — made for the chest, opened it quick, rummaged all through it, more'n once, an' then stood up all white an' glarin'. 'D— him,' sez he — I never heard the lad swear before — 'd— him, he has stolen my diamond!' I thought he must be crazy, sir, with that mountain fever, belike, that the miners get in the diggins. 'Why George, lad,' sez I, 'you're jokin' me. How should a poor sailor-boy 'ave a real diamond — leastwise a honest boy like you?' But George he only lowered at me, an' rushed for the door. He was off into the darkness an' fog before I could stop him, an' though I looked an' called after the lad, I could n't find him. Next mornin', when I opened the bar early, I seen a crowd standin' beyond there, sir, nigh the end o' the wharf. A man comin' from it told me a drownded body was fished up there. Mistrustin' suthin', sir, I went to spy the body. It was the puir lad's! I felt guilty like an' awfu'. I took charge of the body, sir, an' give 'un a good funeral at Yerba Buena. Next mornin' the "Alta" said as how a young man from the mines 'ad fallen through a mantrap in the wharf, 'an give his name as they had it from I. But, sir, whether that be so, or he jumped off mad into the water, seekin' death willfully, I dunno; but I have my thoughts. I wrote to his old mother in England all about his end; but it was a sad job, sir."

The good fellow's voice grew husky as he spoke. I could not speak myself for a few minutes — poor George's fate seemed so sad. Who could have believed that a pure delusion would lead one ignorant man to a mean crime, profitless as he found it, and another to frenzy and death! Who would have suspected such a tragic sequel to our careless chat on the Trinity!

THE OLD AND THE NEW.

In no white winding-sheet goes out the year,
Stiff, straight, and cold, with mourners by its bier,
    As in the hard Atlantic clime,
Where bare-branched trees make desolate the sky,
And streams are stilled but winds are piping high,
    And vapors turn to stinging rime.

Not typical of death our old year's end,
But rather like the parting of a friend
    Who leaves a grateful sense behind;
Or like a maiden loved and wedded late,
Who goes to meet her joy with mien sedate,
    Yet calmly happy in her mind.

The long dry summer sits upon the hills
In memory yet; her russet color fills
    The distant scene with mellow tints;

Only the spring that swells to meet the cloud,
Or acorn-dropping oak, or south wind loud,
    Another mood of nature hints.

The red geranium gleams along the wall,
The pea-vines leafy tresses thickly fall,
    While roses blush in open air;
And oft in sheltered spots, 'mid friendly calms,
The calla lily lifts its broad green palms
    And blossoms into saintly prayer.

Soon all the tawny hills that thirst for rain
Will don an emerald robe with golden train
    Of yellow poppies glowing like a flame;
The summer from her dusty chrysalis
Will waken to a life of wingéd bliss,
    And spring will be its happy name.

www.ingramcontent.com/pod-product-compliance
Lightning Source LLC
Chambersburg PA
CBHW020322240426
43673CB00039B/894